Biomass Compaction

Peter Križan

Biomass Compaction

The Effects of Pressing Chamber Design Parameters on Extrusion Quality

Springer

Peter Križan
Faculty of Mechanical Engineering
Slovak University of Technology in Bratislava
Bratislava, Slovakia

ISBN 978-3-030-89955-4 ISBN 978-3-030-89956-1 (eBook)
https://doi.org/10.1007/978-3-030-89956-1

This Springer imprint is published by the registered company Springer Nature Switzerland AG
The registered company address is: Gewerbestrasse 11, 6330 Cham, Switzerland

Preface

Our institution (Institute of Manufacturing Systems, Environmental Technology and Quality Management (ISETM) in the Faculty of Mechanical Engineering, STU, Bratislava) has long been involved in the scientific development of leading technologies in the manufacturing technology and energy sectors, primarily design, research, and development of pressing machines, which are used to extract wood waste from phytomass and convert them into a useful form of solid biofuel. Some tasks concentrated on the comprehensive design of production lines for the recovery of waste, while others concentrated on the development of pressing machinery—pellet mills or briquette presses or, more precisely, individual parts of these machines (tooling, chamber design, loading, cooling, etc.). The field of recovering energy from various types of biomass waste by means of compaction is a problem which is directly related to ecology and environmental protection.

Based on ever increasing demands for energy independence, and the fact that useable sources of material are diminishing with the requirements for effective extraction, a higher level of research and development within this field is more necessary than before. Currently, it is no longer sufficient to simply manufacture solid fuels, it is now necessary to produce it more cost effectively and with materials which were never initially intended for compaction. Based on personal experience, manufacturers of pressing machines and producers of solid biofuels dedicate relatively little attention to the scientific knowledge of the pressing process, leading to thought processes that may not yield the desired effect. A thorough knowledge of pressing technology, their parameters and their effect on the resulting product, whether in pellet or briquette production, leads to appropriate design and sizing for successful operation of production lines. In view of the above, it is my opinion that the problems are very relevant and beneficial to society.

Biomass Compaction: The Effects of Pressing Chamber Design Parameters on Extrusion Quality is based on many years of experience in the field of experimental research in compaction processes for various types of biomass as well as knowledge gained through various projects and implementations. In numerous cases, I have met with industrial requirements for the optimization of the pressing process, or more specifically, machines for different types of materials. On top of this, renowned manufacturers of pressing machines, for example world-famous German or Danish companies, are beginning to understand the importance of experimental research within the pressing process and the optimization of machines in respect to this as they have personally reached out to our organization for cooperation. Based on this experience I consider my activities in this area to be justified and beneficial to reduce production costs of solid biofuels, thereby streamlining the recovery of various types of biomass and waste.

The main task of this book is to provide a set of information related to the recovery of wood waste by means of pressing and to strive for the most detailed description of the pressing process in order to gain knowledge on the effects of the machine and selecting suitable design parameters of the pressing chamber and their effects on the final quality of extrusions. This book is directed primarily to students in secondary (MSc) and tertiary (PhD) study, but also to interested persons. The presented scientific book aims to offer students of production, environmental and energy technologies and students of mechanical engineering a general understanding of the influences of important parameters in the process of wood biomass compaction. This book can also be useful for experts and/or technicians who deal with the production of solid biofuels (pellets, briquettes) and the construction of equipment for the production of solid biofuels as well as experimental research in the field of compaction.

Extensive and long-term research activities at our institution have resulted in valuable knowledge gained about the influence of important technological and material parameters on the resulting quality of extrusions within the pressing process. Currently, it is possible to accurately determine and define the effects of pressing pressure, pressing temperature, moisture of the input material, and the size of the input fraction on the resulting physical and mechanical properties of the extruded products. The aim of this book is to summarize knowledge, results, and personal experience regarding the effects design parameters of pressing machines and pressing chambers. Therefore, the book is divided into several connected parts. The introduction provides an overview of commonly available design solutions for pressing machines and their basic differences. Although there are several types of pressing mills, their common feature is the working principle. The input material is compressed, by means of a pressing tool, into a compact extrusion due to the actions of pressure and temperature. The individual basic differences in construction of pressing machines result in different operating and production costs, but also differing quality of product. The introduction is followed by a chapter dedicated to the theoretical analysis on the effects of machine construction on the quality of extrusion and analytical analysis on the effects of pressing chamber parameters on the quality and coherence of the extrusion. The goal of this chapter is to identify the importance of the pressing method, type, and shape of the pressing tool.

The content of the following chapter is the shape and dimensional analysis of the extrusions. From the consumer's point of view, the shape and size of the extrusion is an important criterion, but from the point of view of the pressing machine manufacturer and biofuel producer, it is a parameter that significantly affects production, operation, and cost of the entire production process. The shape and size of the extrusions is a parameter that also affects the burning process of the product, automation of the combustion process, storage and transport, and wear of the pressing tool. All these aspects are associated with certain financial requirements, and therefore consider it important to provide even this "look" in harnessing biomass and waste. All these aspects are associated with a certain financial requirement, which is therefore important to consider when recovering energy from biomass and waste. The fifth and sixth chapters form the core of this book and are also the most interconnected. The fifth chapter deals with the theoretical analysis of the influence of select design parameters of the pressing chamber, which is an important and necessary part of experimental research. It discusses and demonstrates the dependencies and importance of design parameters of the pressing chamber with regard to the resulting quality of extrusions. The influence of the pressing chamber length, diameter, and conicity, coefficient of friction, and shape of the pressing tool are analysed in detail. The sixth chapter, which is directly associated with the previous one, describes the implemented experimental research, from setting the research goals, design, and choice of experimental conditions to the presentation of results from the implemented experimental research. Following the chapters which deal with the theoretical analysis and experimental research of influencing design parameters of the pressing chamber, the final part of the book summarizes the results and significance of the experimental research, practical use of experimental research results, and possibilities for further work in this field.

The goal of this book is to summarize knowledge and personal experiences developed in the field of wood waste compaction while simultaneously sharing the results of analyses and experiments carried out at our institution.

I would like to personally express thanks to the professional reviewers and all friends and colleagues who have endeavoured with me on solving problems and tasks in the field of compaction, as well as colleagues from the Faculty of Mechanical Engineering at STU in Bratislava, with whom I've participated in solving many projects of value which have contributed to this book, the field of biomass compaction, design of production lines, and constructions of pressing machines and equipment for the mechanical treatment of biomass.

Finally, I would like to express my sincere gratitude to my wife Lýdia, and sons Viliam and Kristián together with my parents for their unwavering support, understanding, and patience while providing me the conditions to pursue my scientific work.

Bratislava, Slovakia Peter Križan
July 2021

Acknowledgements

Biomass Compaction: The Effects of Pressing Chamber Design Parameters on Extrusion Quality is based on research results from the project VEGA 1/0420/16—"Determination and research of parameters influence on final briquettes quality at waste biomass densification" which was financed by the Ministry of Education, Science, Research and Sport of the Slovak Republic and the Slovak Academy of Sciences, and from the project APVV-18-0505 "Development of the original design of the briquetting screw press with the inverted kinematics" funded by the Slovak Research and Development Agency.

About This Book

This book discusses the scientific process of biomass compaction, mainly the research of pressing chamber parameters and their influence on the quality of extrusions from biomass. This scientific book, a monothematic work, yields new knowledge in the field of wood biomass pressing technology. This book contains a thorough analysis and detailed theoretical analysis of the pressing chamber of pressing machines and the influence they have on the resulting quality of extrusions. Part of this book is dedicated to the proposal and evaluation of experimental research dealing with the definition of different pressing chamber parameters in pressing machines and their effects on the quality of extrusions. Definition and specification of the dependencies of chamber parameters based on the resulting quality of extrusion, given by the mechanical indicators of quality, are also explored. Furthermore, the work describes the design and manufacture of an experimental pressing stand, which allows for experiments to be performed determining the effects that some technological, material, and construction parameters have on the resulting quality of extrusions. The desired pressing method, length, and conicity of the pressing chamber are experimentally determined by means of the uniaxial compaction of wood biomass where results and dependencies are expressed graphically. The significance and practical use of the research results are also discussed within this book, while further recommendations for research in this area are also considered.

Contents

List of Used Symbols and Abbreviations

α	Taper (conicity) of pressing chamber wall (°)
α_v	Excess air factor (–)
AR	Pellet abrasion (%)
BL	Briquette press
D_k	Diameter of the pressing chamber (opening) (mm)
D	Diameter of the pressing chamber (mm)
d_n	Diameter of the extrusion (mm)
D_0	Diffusion coefficient at 0 °C ($m^2.h^{-1}$)
D_p	Mechanical resistance of pellet (%)
ε	Porosity (–)
f_c	Strength of extrusion (Pa, MPa)
F_m	Pressing force on the plug (N)
F_V	Axial force to eject the extrusion from the pressing chamber (N)
F_μ	Friction force (N)
g	Gravitational constant ($m.s^{-2}$)
G	The ratio between the modules of elasticity in the longitudinal and radial directions of a bundle of fibres
H	Length of pressed plug (mm)
HMB	Top dead centre
L	Length of a pressed column within the pressing chamber (mm)
l_k	Length of part on the pressing chamber (mm)
L_k	Overall length of the pressing chamber (mm)
L_N	Length of extrusion according to standards for biofuels (mm)
l_h	Length of part of the pressing chamber, where some conicity is present (mm)
l_z	Length of the pressed column of plugs (mm)
p	Pressure (Pa, MPa)
p_G	Backpressure in the press housing (Pa, MPa)
p_k	Axial pressure on the piston (Pa, MPa)
p_m	Axial pressure on the plug (Pa, MPa)
p_{m0}	Initial pressure (Pa, MPa)

PL	Pellet press
p_N	Normal pressure (Pa, MPa)
p_r	Radial pressure (Pa, MPa)
$p_{r,r}$	Residual radial pressure (Pa, MPa)
P_V	Axial pressure to eject the extrusion from the pressing chamber (Pa, MPa)
r	Radius of the pressing chamber (opening) (mm)
r_z	Characteristic dimension of the extrusion (mm)
S_{PL}	Surface content of a given element of a conical pressing chamber (mm^2)
S_{V1}	Upper surface content of a given element of a conical chamber (mm^2)
S_{V2}	Lower surface content of a given element of a conical chamber (mm^2)
T	Pressing temperature (°C)
τ_h	Nusselt's criterion (–)
T_m	Core temperature of the flame (°K)
ISETM	Institute of Manufacturing Systems, Environmental Technology and Quality Management
μ	friction coefficient
ν_{LR}	Poisson's constant for a bundle of fibres in the longitudinal directions
ν_{RL}	Poisson's constant for a bundle of fibres in the radial direction
V_ε	Volume of gaps (mm^3)
$V_{V\min}$	Volumetric consumption of air ($m^3.kg^{-1}$)
V_t	Volume of the extrusion (mm^3)
ρ	Density of wood ($kg.m^{-3}$)
ρ_0	Density of completely dry wood ($kg.m^{-3}$)
ρ_n, ρ_N	Fuel density of the extrusions ($kg.m^{-3}$)
σ_0	Initial stress (Pa, MPa)
σ_a	Axial stress (Pa, MPa)
σ_m	Axial stress on the plug (Pa, MPa)
σ_p	Compressive strength of the extrusion (Pa, MPa)
σ_r	Radial stress (Pa, MPa)
$\sigma_{r,r}$	Residual radial stress (Pa, MPa)
σ_s	Middle stress (Pa, MPa)
σ_t	Tensile stress (Pa, MPa)

Indexes and remaining denotations are described in the text

Chapter 1
Introduction

Evaluating the energy density of renewable energy sources (RESs) is an inseparably part of modern industry as society continues to solve the problem of ever-growing demands on energy. The potential of RES is determined by the amount of energy that can be extracted or converted from a given raw material, which largely depends on its natural properties. One RES that has displayed great potential in this respect is biomass; however, as the energy demands of the European region continue to rise, their use is still underutilized. Biomass is a suitable source for the production of biofuels. Currently, first-generation biofuels are already available; however, ever-improving technologies are now processing second-generation biofuels, which are assumed to replace a large portion of fossil fuels utilized in the transportation industry. This new-generation biofuel will further reduce greenhouse gasses while having better qualitative parameters. Continued development and use of biomass depend on the cooperation between the environmental, mechanical, and power engineering disciplines supported by research and development, which will lead to the optimization of technologies with respect to their energy effectiveness. The environment provides many sources, which engineers can extract and convert into a desired type of energy. A typical example is the conversion of biomass into solid fuel, which can then be used as heat energy in power plants or other direct heating applications. It is, however, important to utilize natural sources with respect to their effectiveness, economy of production, and without damaging the environment. It is therefore important to extensively analyze all dependent processes and strive to optimize them based on the required criteria. The energy requirements of the European countries, however, cannot be solved by the conversion and use of biomass alone (along with all the dependent technologies); therefore, its widespread implementation is not reasonable.

Recently, more emphasis is being put on the extraction of energy from alternative sources of biomass such as waste from wood. The mechanical processes necessary for its preparation require the refinement of the input material by homogenizing the size and composition of the waste material, adjusting its moisture content, and

P. Križan, *Biomass Compaction*, https://doi.org/10.1007/978-3-030-89956-1_1

controlling other physical–mechanical properties. By refining the input material, it is possible to optimize properties for burning (duration and evenness of burn, shape for automated manipulation and transport, chemical properties, etc.). Such refinement has unlimited stability without biodegradation, provided some conditions are met. For the aforementioned reasons, it is a justifiable process, and moreover, it is necessary to ensure a harmonious relationship between energy extraction from waste and protecting the environment, particularly with the use of best available techniques (BATs). Furthermore, the process utilizes alternative energy sources more efficiently, effectively, and reliably without further impact on the environment. For the extraction of biomass in this respect, pressing technology is discussed, which is fundamental in the "chain" of biomass preparation: biomass ➔ refinement technology ➔ conversion technology; see Fig. 1.1.

Pressing technologies are one of the possible solutions in how to transform suitable organic raw material into modern biofuel. Currently, one of the more known technologies for the compaction of materials is briquette and pellet extrusion. The production of solid biofuels is a suitable treatment method for effective energy extraction from biofuels. Twenty-first-century fuels must satisfy high comfort and safety criteria as well as fulfill environmental and economic requirements. Modern energy carriers must be made from safe, low-value raw materials—waste, which has a homogeneous fraction size, density, moisture content, and shape. The use of pressing technologies in the process of biomass recovery is a deciding aspect in the efficient recovery of biomass and the achievement of higher fuel quality (see Fig. 1.2). As can be observed in the figure, the effects that significantly affect the final quality of the briquette/pellet are apparent at the stage of raw material processing and, of course, also depend on the type of the input material.

Many projects dealing with extracting, conversion, and processing of wood waste and waste from phytomass by means of pressing have been investigated at the Institute of Manufacturing Systems, Environmental Technology and Quality Management (ISETM), Faculty of Mechanical Engineering, Slovak University of

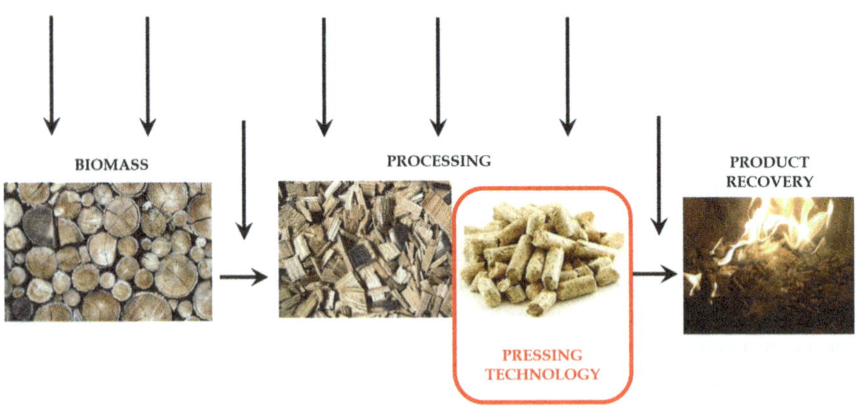

Fig. 1.1 Role of pressing technologies in the extraction of biomass [1]

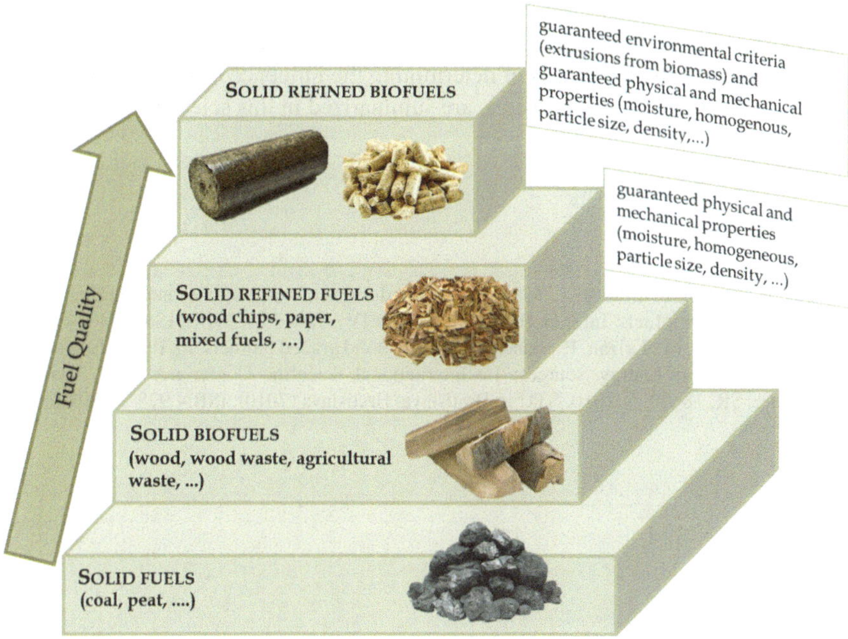

Fig. 1.2 Quality of solid fuels with respect to the level of mechanical refinement [2]

Technology (STU), in Bratislava, Slovakia. During these extensive research investigations, valuable information regarding the impact of various technological and material parameters on final quality of product has been determined. As a result, it is now possible to determine the effects of chamber pressure, chamber temperature, moisture content, and fraction size of the input material on the resulting physical and mechanical properties of the briquette/pellet. It is, however, important to not overlook the structural parameters of the pressing machine itself, particularly the pressing chamber, and their effects on the resulting quality. In practice, many cases were encountered where "improvements" to the pressing process with the intention of improving the physical and mechanical parameters of the resulting product were limited by the pressing machine itself. In these cases, it is necessary to optimize the parameters of the pressing machine, particularly those of the pressing chamber.

This scientific monograph deals with the construction parameters of pressing machines and their effects on the quality of the resulting solid biofuel. The function of pressing technology is to produce a solid compact biofuel from suitably refined input material that fulfills the necessary qualitative properties. The market has various types and constructions of pressing machines. With technological development and experimental research, it is possible to manufacture a pressing machine with higher effectiveness in the production of solid biofuels. In the design of new pressing machines, or improvement of existing machines, it is necessary to have a thorough understanding of the pressing process, its parameters, and their effects on the extruded briquette or pellet. The quality of the extruded product is highly dependent on the composition of the input material. Just as important are the methods and construction parameters of the machine. Some results of my scientific research activities and efforts in the field of determining the influence of individual parameters on the final quality of extrusions are summarized in this scientific monograph.

References

1. Šooš, Ľ., Križan, P., Beniak, J., Matúš, M.: New designs of compacting machines and shapes of pressing bio-biofuels. Int. Sci. Index. **17**(10, part IV), 550–555 (2015). ISSN 1307-6892
2. Šooš, Ľ., Matúš, M., Križan, P.: Standardization of solid graded biofuels. In: Power Engineering 2010. Renewable Energy Sources: 1st International Scientific Conference OZE. Tatranské Matliare, SR, 18–20. 5. 2010. STU in Bratislava, Bratislava (2010). ISBN 978-80-89402-24-3 (in Slovak)

Chapter 2
Construction and Types of Pressing Machines

Pressing technologies are one of many solutions in transforming suitable organic raw material into a modern biofuel. One of the more well-known pressing technologies is the pressing of briquette pellets and compaction. The primary difference between the aforementioned technologies is the shape and size of the pressed product, and the process involved in its formation. The pressed product is then both materially and energetically ready for consumption. Compaction is primarily used with nonbiological material (metal shavings, metallic dust, coal, chemical dust, tablets) since the extrusions typically have difficultly retaining their integrity over time. For this reason, this monograph only deals with the technology used for pressing briquettes and pellets, both of which are similar in process where high pressures and temperatures are in effect. Briquette and/or pellet pressing also rarely require a binder to retain integrity of the final extrusion. Based on the composition and type of pressed material, cell structures within release lignin when subject to certain temperatures and pressures. This, in combination with sufficient pressing and slow cooling of the extrusion, acts as a binder.

Briquette products are typically cylindrical, cuboid, or n-angled shape, which can be burned easily in open combustion equipment. In briquette formation, one briquette is extruded per pressing as opposed to a pellet extrusion. Products available for the production of briquettes from waste materials are rather broad. Briquettes can be made from pressing various materials such as sawdust, wood chips, bark, wood shavings, straw, paper, cotton, and waste from mineral resources. During the briquette pressing process, the surface "melts," which results in a stronger, more resistant extrusion even without the use of binders or other additives. This property makes manipulation of the briquette easy and dust-free. Briquettes also have an advantage in duration of combustion and the ability to be used in virtually any open system. Common uses are in stoves, fireplaces, central heating, wood boilers, etc. Although they are unsuitable for small- or medium-sized automated combustion processes, common briquette machinery are either mechanical (utilizing a crank, linkage, or screw mechanism) or hydraulic (utilizing a hydraulic piston

P. Križan, *Biomass Compaction*, https://doi.org/10.1007/978-3-030-89956-1_2

dimensioned according to the desired size and type of material to be extruded). The following Figs. 2.1, 2.2, 2.3, 2.4, 2.5, 2.6, 2.7, 2.8, and 2.9 display the current design and principles of common briquette presses, where the pressing tool is either a piston or screw mechanism.

Many worldwide manufacturers of briquette presses are beginning to understand the need for experimental research of the briquette pressing process necessary to obtain optimum efficiency and use from these machines. Despite this, they seem to underutilize the scientific knowledge of the process itself and commonly design their machines around the concept of increased production by the simple addition of chambers to a single machine; see the following Figs. 2.10 and 2.11.

Pellet production is the process of compressing crushed and dried material through an extrusion, which typically has multiple openings. It is characterized by the production of numerous extrusions, pellets at once. After extrusion and parting, the pellets are very hot and plastic. Only after cooling do they harden and achieve some mechanical resistance. This technology is demanding on the quality of the input fraction, which must be finer than that used in the briquette process. Pelletization can be used for processing various types of waste, such as straw, pulp, chips, bark, coal dust, waste from minerals, food products, fertilizers, and PVC waste and is particularly useful in the pharmaceutical industry. Currently, the market offers many types of pellet machinery. The following Figs. 2.12, 2.13, 2.14, 2.15, 2.16, 2.17, 2.18, 2.19, 2.20, 2.21, 2.22, 2.23, 2.24, 2.25, and 2.26 display the well-known design and principles of pelleting presses. They can be manufactured with either a mechanical drive or a hydraulic drive. The pressing tool may be a rotor, gears, extrusion die, and/or pressing rollers. The extrusion die may be a flat circular surface or cylindrical block, which rotates around either the vertical or horizontal axis. Once the material is extruded through the die, a knife, located somewhere around the circumference of the die, will typically part the pellets at a desired length. The quality of the pellet press is not so much dependent on the principles of the design, but rather on the actual build quality of the machine.

Most manufacturers of pressing mills will concentrate on one type of technology; however, there are manufacturers which will produce both types of

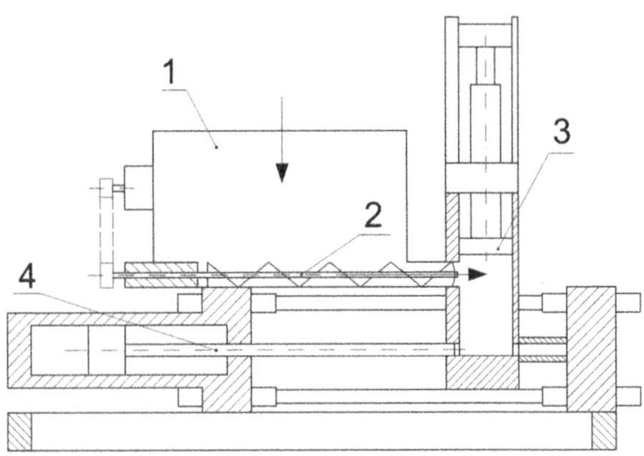

Fig. 2.1 Schematic of a hydraulic press (*1*—feeding, *2*—transport screw, *3*—closing piston, *4*—pressing piston)

Fig. 2.2 Hydraulic briquette press [1]

Fig. 2.3 Hydraulic press for cylindrical briquettes

technologies. Alongside this, most manufacturers will offer technologies for the preparation of biomass raw materials such as crushers, dryers, and separators, as well as post-processing technologies for cooling, wrapping, and the extruded product. Of course, manufacturers also offer complete solutions based on customer requirements. Although to orient on self in this area is not a simple problem.

The foundation for the design of a production line is to select the type of pressing mill to be used. Production of solid biofuels also requires proper technology and machinery for preparation. Suitable selection of technology and machinery within the production line, taking into account the raw material, is the bases for effective functioning of the production line. When determining the suitability of individual pressing technologies, it is necessary to analyze them in terms of required quality of produced extrusions, type of input material, economic and technical requirements, as well as the requirements of the customers. A thorough analysis of the individual parameters of the pressing mill is a necessary step in establishing a properly

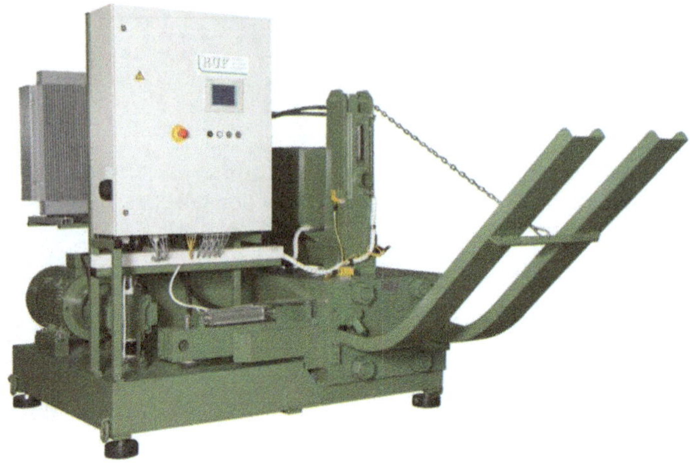

Fig. 2.4 Hydraulic press for rectangular briquettes

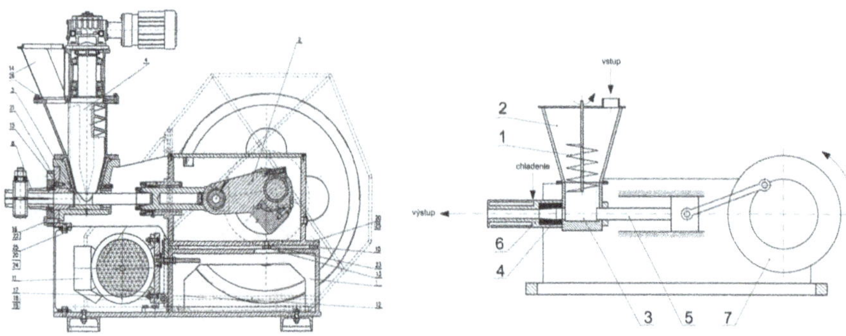

Fig. 2.5 Schematic of a mechanical crank-style press

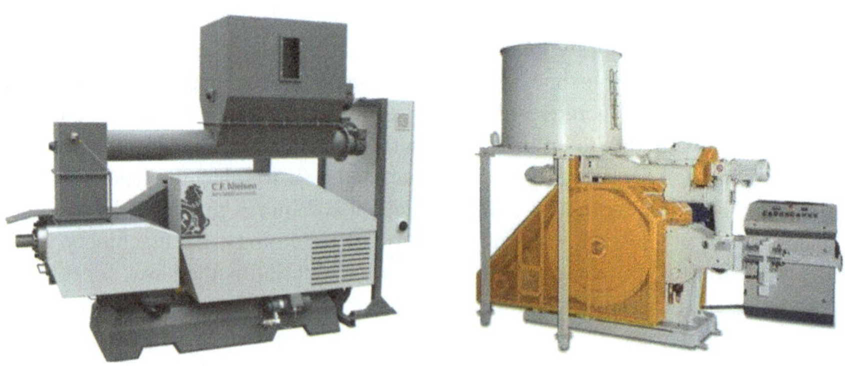

Fig. 2.6 Briquette press based on a crank mechanism [2, 3]

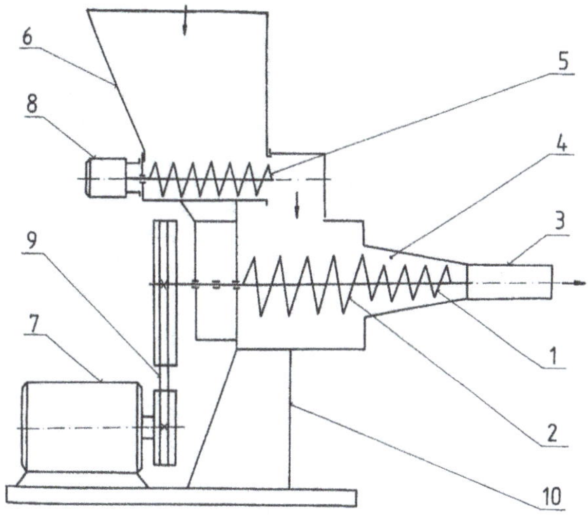

Fig. 2.7 Schematic of a screw press

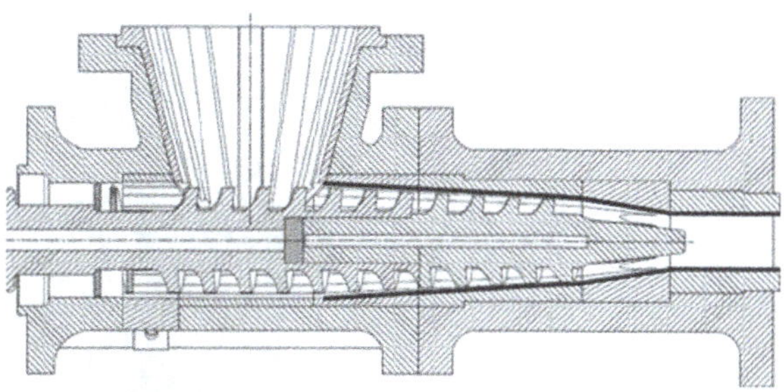

Fig. 2.8 Section of the pressing chamber of a screw press

Fig. 2.9 Briquette press with a screw chamber

functioning methodology for the selection of a suitable pelletizing or briquetting machine, primarily the power requirements of the machine and desired hourly output. The methodology for selecting a pressing mill requires knowledge of the market, available products, and analysis of pressing machines.

Fig. 2.10 Dual-chamber briquette press

Fig. 2.11 Multichamber briquette press [2]

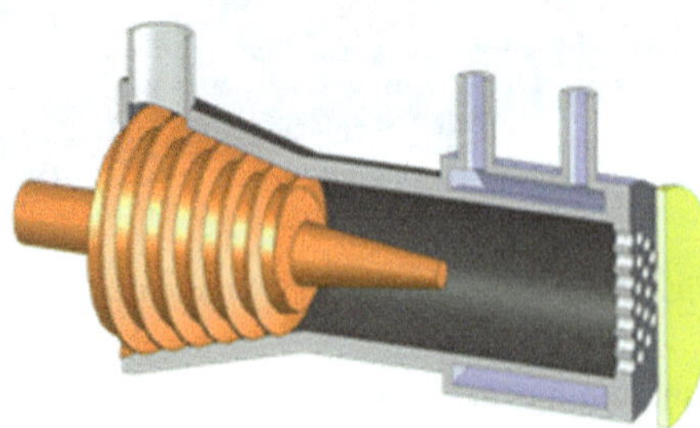

Fig. 2.12 Schematic of a screw-type pellet press

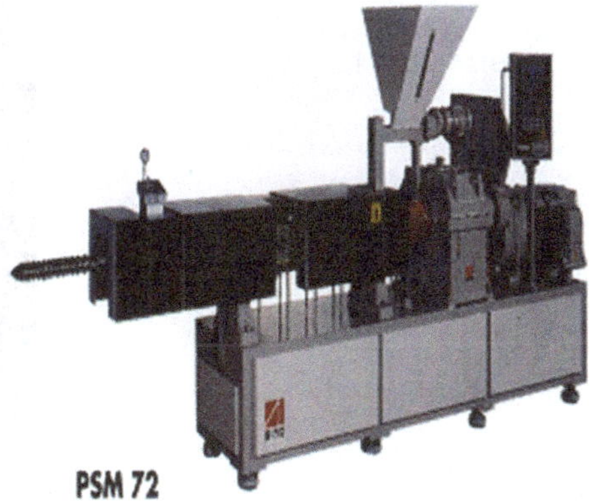

Fig. 2.13 Pellet screw press

Fig. 2.14 Pellet screw press with details on the extrusion die and parting tool

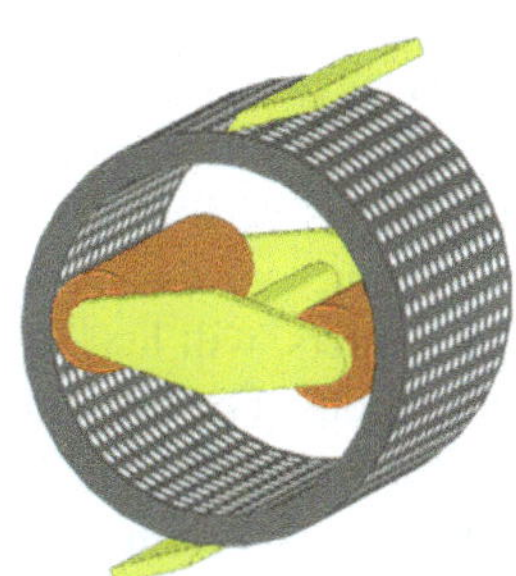

Fig. 2.15 Press with cylindrical die and internal rollers

Fig. 2.16 Press with horizontal cylindrical die and internal rollers [4]

Fig. 2.17 Press with flat circular die and rollers

Fig. 2.18 Pellet press from the manufacturer Amandus Kahl (flat circular die with rollers) [5]

Since this chapter deals with the overview of pressing mills, it goes without saying that all of the aforementioned designs for briquette and pellet mills are offered in various performance ranges. Thorough analysis of each individual design and acceptance of the basic conditions for compaction, with respect to the machines parameters, will lead to the selection of the optimum pressing mill for the required conditions. For illustration, it is possible to show some partial results of analysis performed at the Institute of Manufacturing Systems, Environmental Technology and Quality Management (ISETM); see the following Figs. 2.27 and 2.28.

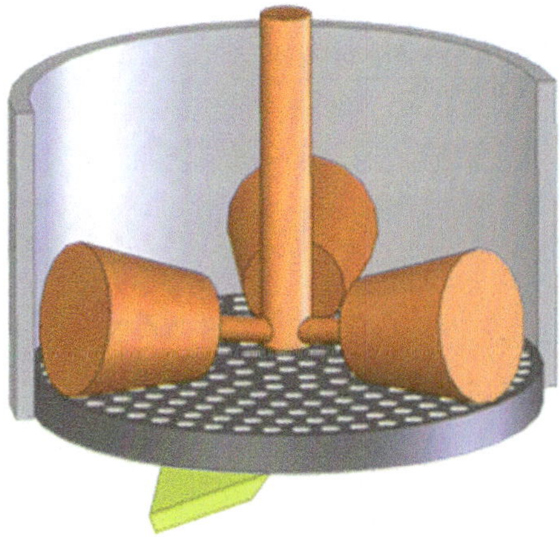

Fig. 2.19 Press with circular flat die with conical rollers

Fig. 2.20 Press with circular flat die and conical rollers

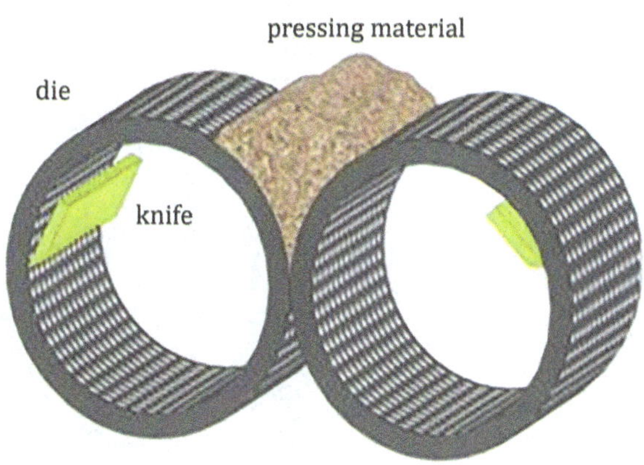

Fig. 2.21 Press with opposing cylindrical dies

The parameters of the analyzed machines and their allocation, according to individual criteria, are the basis of the functionality of the algorithm when selecting a specific machine according to required parameters. Technical information and

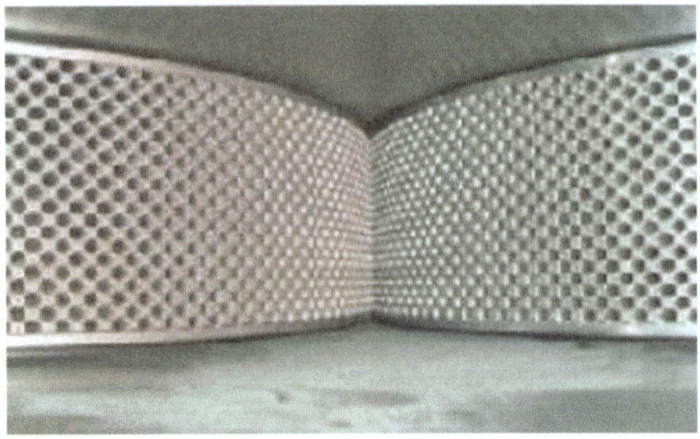

Fig. 2.22 Pellet press by Eco-Tre system (opposing cylindrical dies)

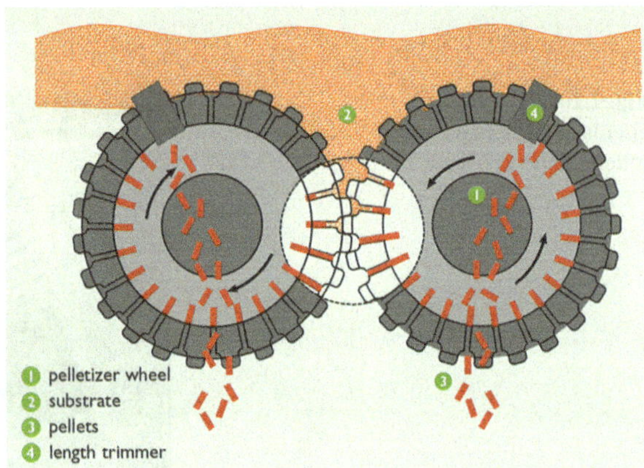

Fig. 2.23 Horizontal opposing gear press

Fig. 2.24 Horizontal opposing gear pellet press

parameters are necessary assumptions in determining the technological and investment costs per unit, which are important in the selection of a particular type of pressing mill and can be directly incorporated within the methodology when selecting a machine. The following Figs. 2.29 and 2.30 show how the analyzed machine parameters apply to determining the suitability for different categories.

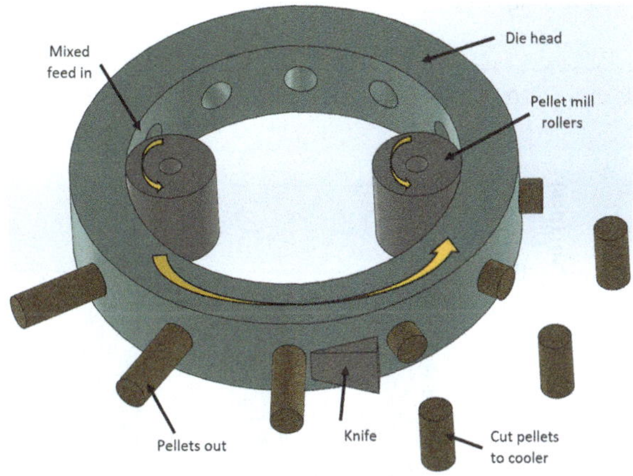

Fig. 2.25 Vertical cylindrical die with internal rollers

Fig. 2.26 Press with vertical cylindrical die and internal rollers

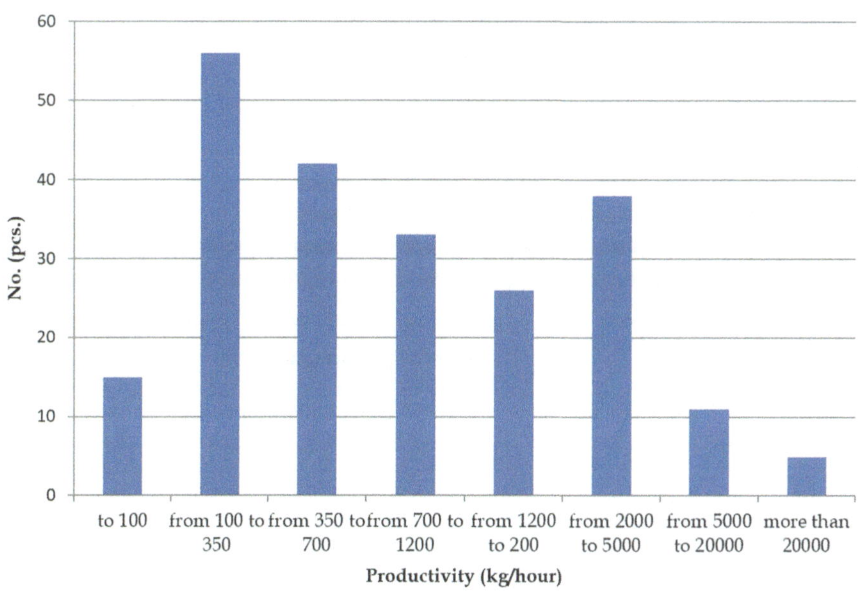

Fig. 2.27 Productivity of pellet production based on output [6]

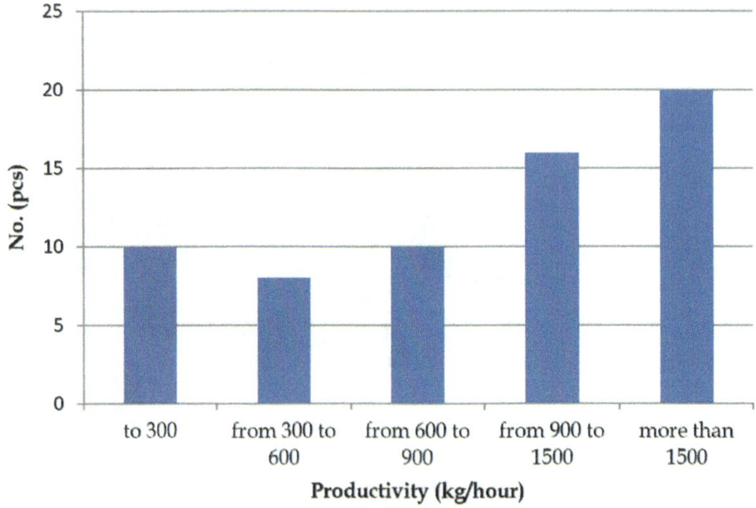

Fig. 2.28 Productivity of briquette production based on output [6]

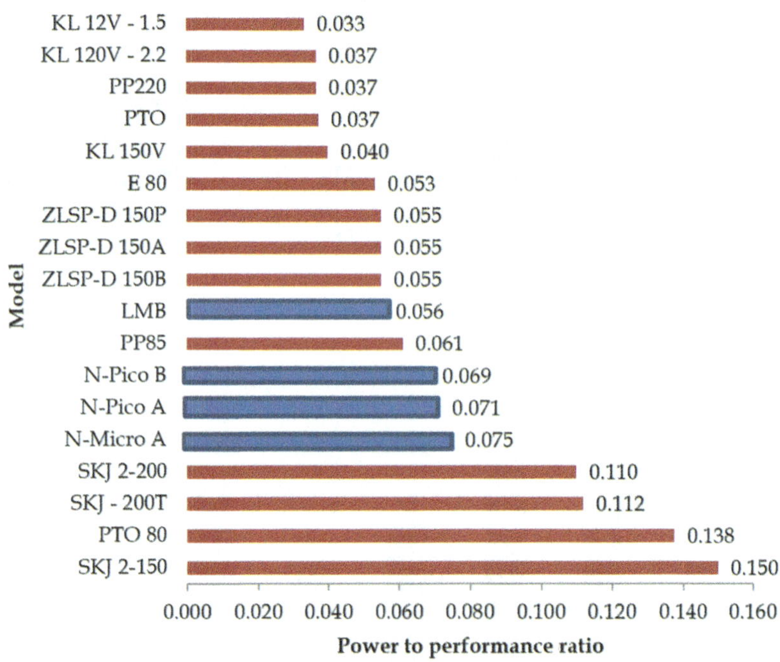

Fig. 2.29 Power-to-performance ratio of different pellet mills with production up to 100 kg/hour [6]

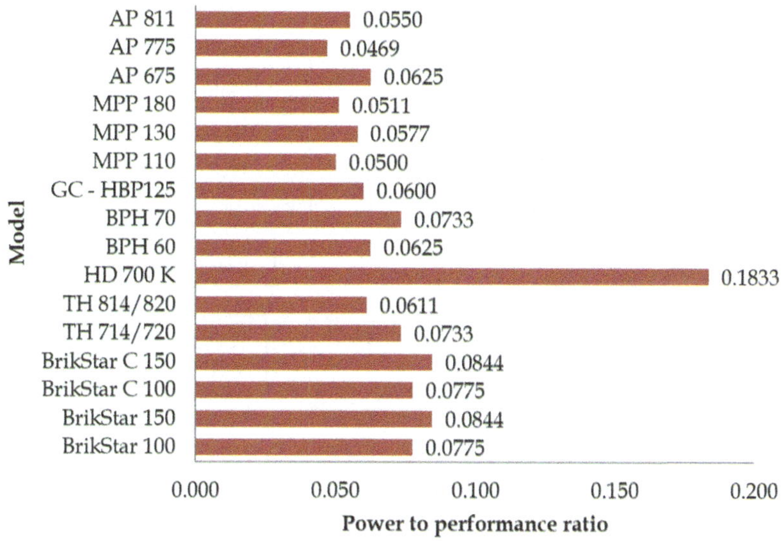

Fig. 2.30 Power-to-performance ratio of hydraulic briquette machines with production from 100 to 200 kg/hour [6]

Afterward, based on the information determined from the thorough analysis of briquette and pellet mills in combination with the computational parameters for the power-to-performance ratio, the technological and investment costs can be established followed then by the design or selection of the most suitable pressing mill, based on initial parameters. When the procedure for the definition of initial requirements is followed, it is possible to then proceed with the process of selecting a suitable pressing machine. Important and deciding initial requirements are:

- Input material type (dendromass, phytomass, etc…)
- Degree of mechanical treatment (moisture content, fraction size)
- Type of extrusion (briquette, pellet, etc.)
- Base dimensions of the extruded product
- Desired hourly production
- Desired quality of production
- Power requirements
- Cost

Such a methodology for the selection of a pressing machine according to specific initial conditions is gradually assessed at each level, which eventually yields a specific pressing mill. Each level of selection consists of defining the type of extrusion, processing method of material, configuration of the machine, hourly production, size of extrusion, definition of the power-to-performance ratio, and overall cost. The procedure for selecting a suitable pellet or briquette mill is performed based on customer requirements, which are given by the input parameters. It is necessary to remark that functionality of the selection process is also defined by a concrete

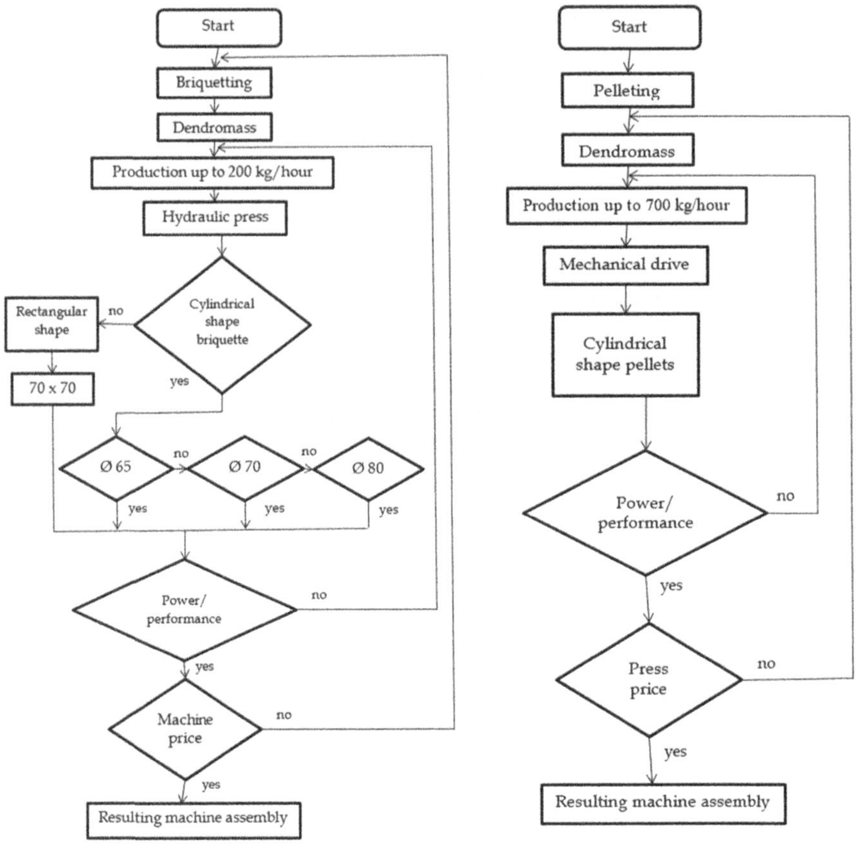

Fig. 2.31 Example of the selection algorithm for briquette machine (left) and pellet machine (right) [6]

definition of the processed material and to, at minimum, conceptualize the production line. The structure of the selection algorithm for briquette or pellet mills takes the form illustrated in Fig. 2.31.

Based on the knowledge summarized in this chapter, concerning the basic design and performance of pressing machines, it is possible to understand the influence of the design, type, and configuration of the pressing machine on the efficiency and effectiveness of the compaction process. Unfortunately, due to diminishing sources of viability material, there is a trend to mix input materials. This fact causes some complications for pressing mill operators. Each type of biological material requires a specific approach and setup of the technology. Since many different pressing technologies exist, there are questions regarding the influence of the design parameters, particularly the pressing chamber, on the quality of the resulting extrusion. This opens up many possibilities for the optimization of the pressing space with regard to the quality of extrusions (briquettes or pellets).

References

1. Briklis, s.r.o.: Company's web page, cited online in April 2021. http://www.briklis.cz/en/
2. C.F. Nielsen A/S: Company's web page, cited online in April 2021. http://www.cfnielsen.com/infoen/-10
3. DI-PIU Macchiné Impianti: Company's web page, cited online in April 2021. http://www.di-piu.com/
4. CPM Europe BV: Company's web page, cited online in April 2021. https://www.cpmeurope.nl/en/products
5. Amandus KAHL GmbH: Company's web page, cited online in April 2021. https://www.akahl.de/en/machines/pelleting/pellet-mill/
6. Šooš, Ľ., Križan, P., Matúš, M.: Solid graded fuels as an alternative to conventional energy carriers. In: Energetika a biomasa seminar, pp. 62–72. ČVUT, Praha (2010) (in Slovak)

Chapter 3
Influence of Machine Design on the Quality of Extrusions

The manufacture of solid biofuels is a suitable way of effectively extracting energy from biomass. Twenty-first-century biofuels, apart from fulfilling environmental and economic criteria, must also be comfortable and safe to use. Several norms [1–4] have been enforced to ensure that the production, sales, and distribution of biofuels conform to a certain level of quality. In general, solid biofuels are considered to be of high quality when fulfilling the aforementioned norms. The technical standards define various parameters (indicators), which are used to evaluate the quality of the solid biofuel. The indicators of quality of solid biofuels can be divided into two basic groups [5]:

- Physical–mechanical indicators
- Chemical–thermal indicators

The treatment of biomass, where compaction is involved, primarily affects the mechanical indicators. According to STN EN ISO 17225-2,3 [6, 7], the physical–mechanical indicators of quality are dimension of extrusion, bulk and specific weight, size of particles, density of particles, and mechanical resistance. When assessing and determining the influence of design parameters on the pressing mill, it is necessary to evaluate results primarily with respect to the aforementioned physical–mechanical indicators of quality. The most complex parameters, which basically take into account all the remaining quality indicators, are the density of extrusions and their mechanical resistance. When defining the mechanical resistance according to relevant technical standards [8, 9], it is necessary to use special laboratory equipment. The density of extrusions is possible to be calculated based on the dimensions and weight of extrusions [1]. The density of biofuel limits the ease of which it can be transported, and also the effectiveness of extracting energy from it, minimizing transportation and storage costs. The chemical–thermal indicators of quality depend mainly on the type and origin of the biomass. This includes the moisture and ash content, heating ability, content of carbon, hydrogen, nitrogen, chloride, sodium, potassium, sulfur, chlorine, and volatile substances [6, 7]. Biomass

P. Križan, *Biomass Compaction*, https://doi.org/10.1007/978-3-030-89956-1_3

refinement can also change the chemical–thermal characteristics of biofuels, but in terms of pressing mill construction, it does not make sense to follow this parameter.

When deciding which pressing technology to use and each individual construction type, it is necessary to subject these technologies and machines to analysis concerning the quality of the produced extrusions and the influence of the pressing chamber design parameters. This part of the monograph is devoted to summarizing the knowledge gained from practice, concerning the influence and differences in design principles of pressing machines and pressing tools.

3.1 Influence of Pressing Method and the Pressing Tool Used

Despite the variety of pressing mills, in principle there are only 2 basic ways of compaction resp. pressing. Compaction can be considered with either an open or closed chamber [5, 10].

The principle of compaction in a closed pressing chamber is based on the pressing of material within an enclosed chamber, i.e., on all sides. The feed into the chamber fills it with material, then closes it hydraulically. The nozzle that forms the extrusion is also hydraulically actuated. Compaction occurs primarily through the pressing tool. The resulting density of extrusions obtained this way rarely exceeds 1 $kg.dm^{-3}$. After filling the chamber, compaction occurs, followed by ejection of the pressed material from the chamber. The individual extrusions do not have the conditions to be interconnected; they have been subjected to pressure and calibration in the pressing chamber for extended periods of time, and therefore, it is not possible to achieve extrusions of higher density—quality. The pressing performance can be improved by increasing the amount of input volume, resulting in longer extrusions. However, with this method of pressing, the density and thus the quality decrease as the length of the extrusion increases. This principle of compaction (see Fig. 3.1) can be seen mainly in hydraulic briquetting presses, which operate on the basis of a certain closed working cycle.

Compaction in an open chamber is a pressing method where, during the continuous supply of raw material into the chamber, the material is pressed into the form of a solid extrusion where they gradually move through the pressing chamber and the pressed material is extruded out of the chamber; see Fig. 3.2. The supply of material can be achieved by various means, such as an auger or simple gravity feed. Compaction of the material is primarily achieved by the pressing tool. This type of compaction is capable of reaching densities required by technical standard [11, 12], which is a minimum of 1.14 $kg.dm^{-3}$. Furthermore, this method provides the possibility of certain variability in the construction of the pressing chamber, so that it is possible to increase the quality of extrusions through different modifications within the pressing chamber.

In continuous supply of raw material, compaction occurs by means of the pressing tool, which is constantly moving within a closed cycle. The material is gradually compacted as it moves through the chamber until its point of release or exit from the

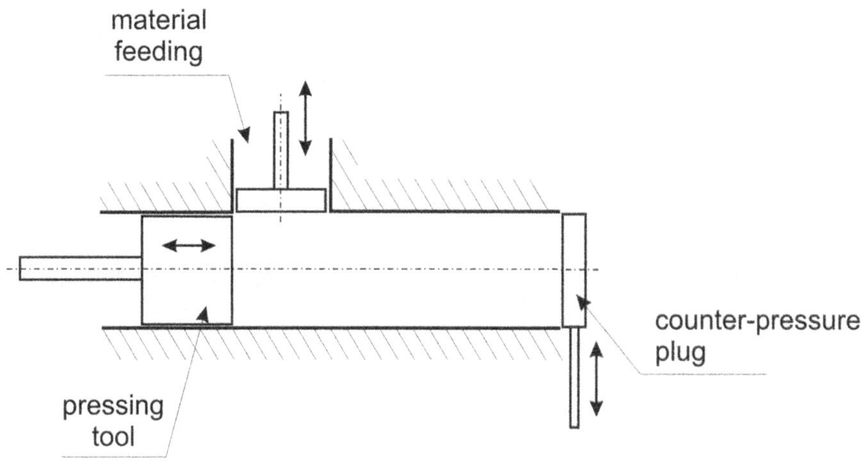

Fig. 3.1 Principle of compaction in a closed pressing chamber

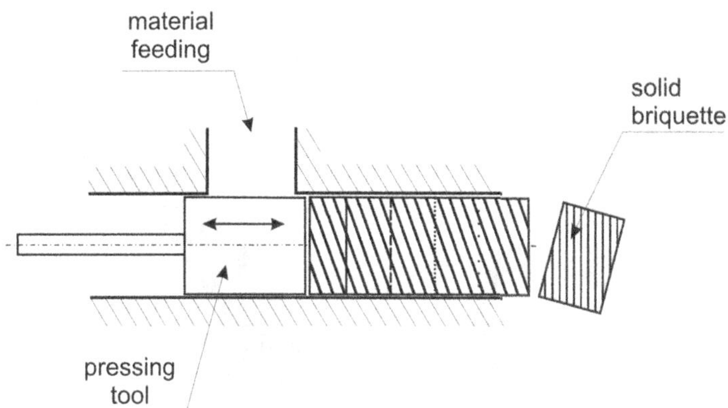

Fig. 3.2 Principles of compaction within an open chamber

chamber. This phase causes the extrusion to undergo prolonged action of the pressing tool pressure (or even temperature). During this phase (otherwise known as the calibration phase), the pressing action releases internal stresses, which would otherwise cause the extrusion to crumble when extruded from the chamber. Also during this phase, conditions are not only suitable for the ideal bonding of material particles within the extrusion but also between individual extrusions. Such properties result in a continuous extrusion that is necessary to part as required. The extrusions are parted either utilizing the overhanging weight of extrusions or mechanically by means of a blade or saw. The aforementioned principles of compaction can be observed in mechanical briquette pressing (by piston or screw tool) or any method of pellet production. Figs. 3.3 and 3.4 show the concept of open chamber pressing, for both plate and cylindrical dies.

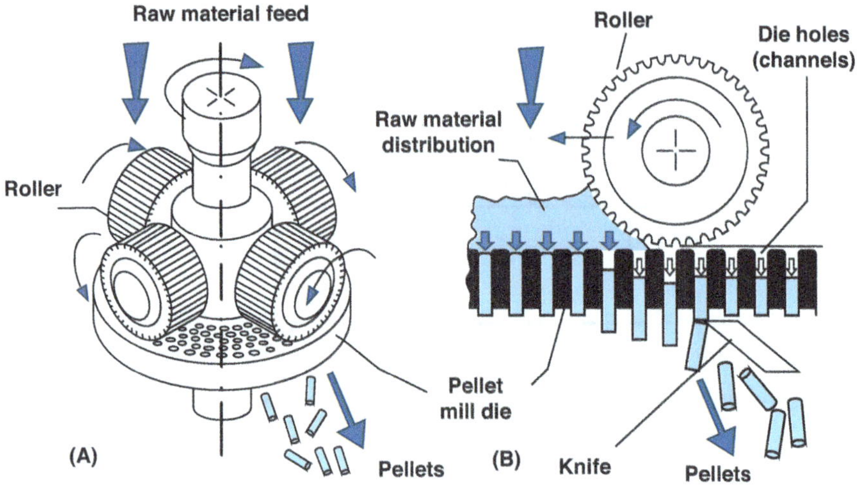

Fig. 3.3 Principles of pellet press with an array of vertical holes within the die [13, 14]

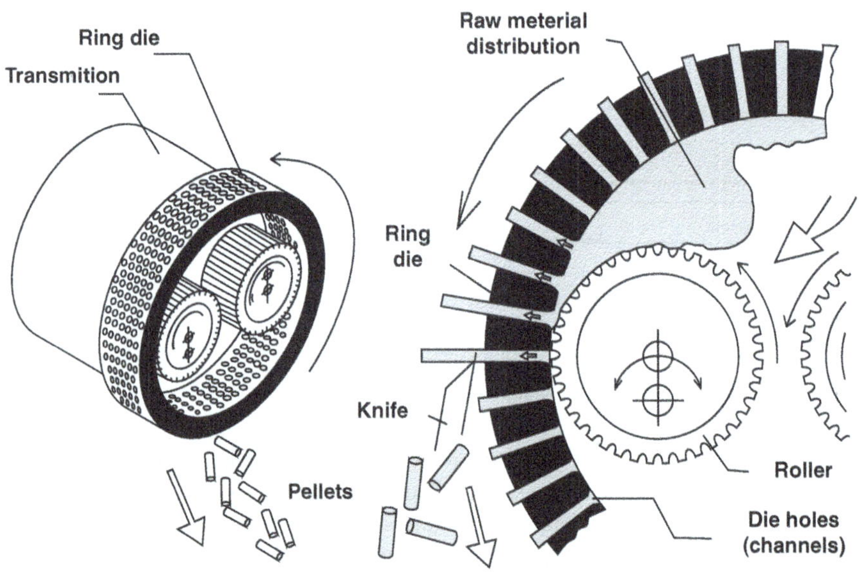

Fig. 3.4 Principle of a horizontal pellet mill utilizing a cylindrical die [11]

When we look closer, we can see that the material is gradually pressed through an array of openings within a die, by means of a rotating roller. This is performed with a constant feed of input material. The material is pressed through the openings where its final shape and density are obtained. Pressing the material through an opening (or array of openings) in this way yields pellets, which are cut at a desired length by a knife. This method of pressing is the same for plate and cylindrical dies.

Their differences are in construction and rotational axis of the pressing nozzle. In a plate configuration, it is the vertical axis of rotation, whereas for cylindrical configurations, it is the horizontal axis. Each requires different ways of transmitting torque from the motor to the pressing tool.

Regardless of the fact that we only differentiate between two basic principles of pressing, there are many construction configurations to achieve either. Therefore, differences in achieved effectiveness and quality of extrusions exist between different configurations. The goal of this monograph is to analyze and declare the effects of the different construction parameters on the resulting quality of extrusions.

The products of extrusions (see Fig. 3.5), depending on the technology used, are briquettes or pellets. Briquettes, based on the type of machine used to produce them, are extruded into cylindrical, rectangular, or n-sided shapes. Pellets, after being pressed through a die and cut to length, are usually only cylindrical in shape. In practice, the most common sizes are 6 and 8 mm in diameter.

As was mentioned before, when producing briquettes, only one unit/extrusion is made at a time (as opposed to pellet production). The pressing tool may be a piston or screw. We discussed mechanical briquette pressing (crank mechanism, lever mechanism, and screw mechanism) and hydraulic briquette pressing. Due to the relatively small size of pellets, it is unnecessary to discuss internal defects with respect to the technology employed. However, for briquette extrusions this becomes much more relevant. Briquettes are larger, and therefore, any internal defects significantly impact the resulting quality of the extrusions (density and mechanical properties of extrusions).

Since many configurations of briquetting presses exist, the question arises, what are the differences in the resulting extrusions (briquettes) with respect to the configuration that produces them? If we look deeper into the processes of compaction in mechanical briquetting presses and compare them with those of hydraulic

Fig. 3.5 Common shapes of briquettes and pellets to date

briquette presses, we can observe some basic differences. In mechanical briquetting presses, the pressed material is always under pressure within the calibration chamber as it is continually fed through the chamber via the axial screw producing a compact briquette over the entire length of the pressing chamber. In hydraulic configurations, compaction is obtained within a closed chamber based on a set stroke of the hydraulic ram and the parameters of the hydraulic circuit. Once the axial pressure is released, the chamber is opened and the compacted briquette is ejected. The calibration phase is limited by the duration of hydraulic ram at full stroke. Therefore, a difference in principle can be observed between the two configurations. For illustration and a better understanding of the differences between the two, a real example can be described. Based on industrial requirements, our institute has performed research on the capability to compact municipal waste with the goal of obtaining a uniform compact product from raw materials with various compositions.

Sometimes, the material may not have an ideal composition with respect to producing bonds between particles within the material during briquette production. In the presented case, the municipal waste was composed of various types of waste, mainly nonbiological materials; see Fig. 3.6. In this case, one possibility to obtain a compact briquette, we can introduce a suitable amount of material containing lignin (biological material). By doing this, the integrity of the produced briquette is much more suitable than without. If the binder is also cheap, then the economic benefits are high as well. Therefore, various materials with binders (wood chips, sawdust, cement, cardboard) were mixed into the municipal waste when testing samples; see Fig. 3.8.

However, another part of this research was to solve the problem by performing the compaction process on different types of briquetting presses (mechanical BP and hydraulic BP). In the following Fig. 3.7, the briquette machines are illustrated, both having a pressing piston and in both instances with a briquette output size of Ø 60 mm (Fig. 3.8).

In Fig. 3.9, we can see the difference between briquette densities from both processes. It is possible to claim that the material composition and type of briquetting technology used significantly influence the density of the produced briquette.

Fig. 3.6 Disintegrated municipal waste (left) and the briquettes made from them (right) [15, 16]

Fig. 3.7 Briquette presses used in research of compacting municipal waste, mechanical crank MP (left), and hydraulic HP (right) [17, 18]

Fig. 3.8 Produced briquettes, from the mechanical (left) and from the hydraulic press (right) [15, 16]

Briquettes made on the mechanical crank press (RDF-MP) obtained a higher density than those produced from the hydraulic press (RDF-HP). For information, the composition of each sample is described:

Sample 1—mix of municipal waste consisting of 38% wood chips from softwood, 45% of shredded cardboard, 11% shredded PET bottles, 6% textile waste (MMW)
Sample 2—municipal waste with an addition of 20% shredded cardboard (80/20—MP)
Sample 3—municipal waste with 4% cement (4C)
Sample 4—municipal waste with 20% sawdust (20 WS)
Sample 5—municipal waste alone (RDF-MP)
Sample 6—municipal waste with 50% shredded cardboard (50/50—HP)
Sample 7—municipal waste alone (RDF-HP)

Figures 3.10 and 3.11 show the strength results of the briquettes in compression. The compressive strength of the briquette was performed by two basic tests—simple pressure test and a splitting test. The briquette is placed between the circular rams of the testing press, where uniform pressure is applied until collapse of the

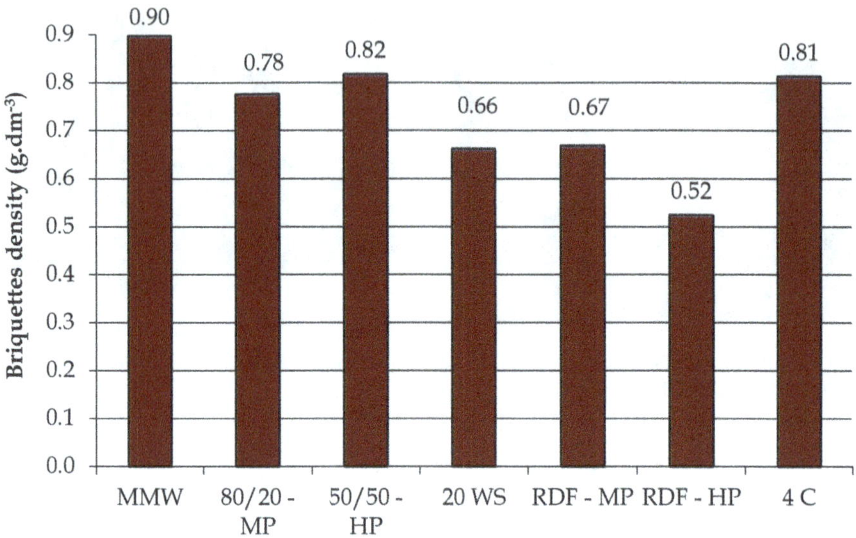

Fig. 3.9 Comparison of briquette density from different compositions of municipal waste [17, 18]

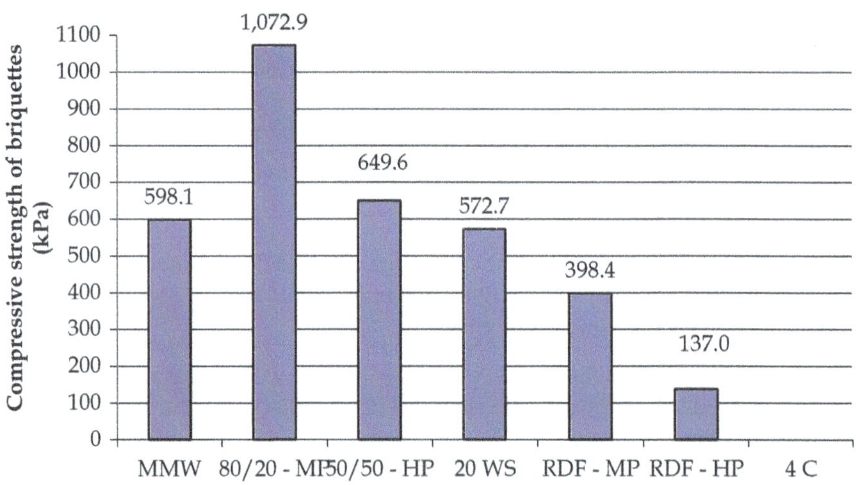

Fig. 3.10 Basic compressive strength of briquette composed of different mixtures of municipal waste [18]

briquette. Tests used only complete, undamaged briquette extrusions. The briquette is subject to gradually increasing pressure until collapse, and the maximum recorded force quantifies the compressive strength of the briquette. The tests displayed the positive effects of an organic binder (paper, wood chips, sawdust, etc.) with respect to the strength of the briquette. These binders assisted with creating internal bonds between the nonorganic particles (PET bottles, textiles, tarps, plastics, etc.).

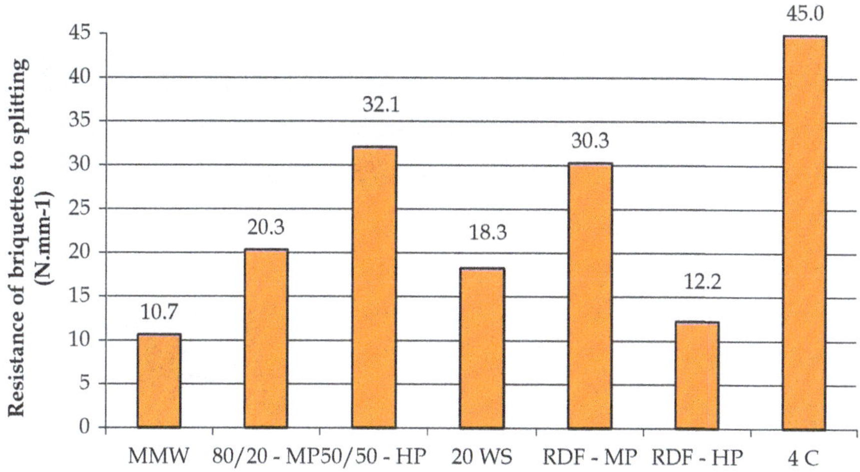

Fig. 3.11 Resistance of different briquette made of mixtures of municipal waste to splitting [18]

Fig. 3.12 Pressing piston tool

Based on the presented results, it is possible to state that, based on the mechanical indicators of quality (density and strength), there is an obvious advantage with the use of a mechanical BP since the "open chamber" compaction process positively influences the creation of bonds between particles. Based on parameters of dimensional and shape precision, the hydraulic BP is more suitable, since the compaction occurs in a "closed chamber" where the briquette is produced individually in a set shape and regardless of the lower mechanical properties, briquettes produced by hydraulic machines have flat faces, which resist crumbling better and have consistent length.

Problems regarding process optimization and resulting quality of extrusion are not always a question of the mechanical configuration. Although the type and principle of pressing process have proven to have a significant effect on the resulting quality of extrusions, the pressing tool used can be divided into three groups, where each principle has its specific effect on the quality of the extruded briquettes. The most commonly used pressing tools are illustrated in Figs. 3.12 and 3.13.

Part of a piston press is usually a crown on the piston head (see Fig. 3.14), which helps to induce shape bonds when pressing the material, which improves the

Fig. 3.13 Pressing screw

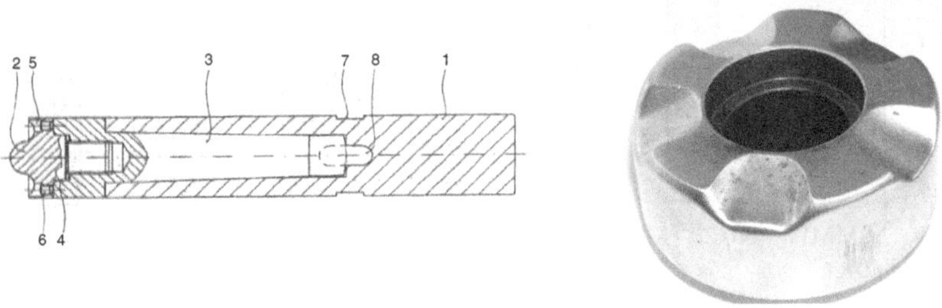

Fig. 3.14 Shaped crown and its place on the pressing piston [19]

integrity of material particles without the use of a binder. This is an important fact to consider as the piston crown often is subject to the most wear.

Considering the character and intention of this monograph, it is important to analyze the type of pressing tool and its effect on the resulting quality of extruded briquettes. It is necessary to state that the published differences and analysis results are valid only for briquette tools since the compaction process in pellet mills, pressing material through an array of openings by means of a rolling element, is essentially all the same. This is not to say that the pressing chamber cannot be optimized to affect the resulting quality of extrusions, but the shape of the rolling elements does not significantly affect the quality of extrusions. For pellet mills, the significant parameter is the geometry of the openings within the die through which the material is pressed, which to an extent can be compared to the effects of pressing chamber geometry in briquetting presses. However, this effect is much more enhanced within briquette presses and also affected by the shape of the pressing tool.

Pressing piston of a hydraulic press—when producing briquettes from biomass on a hydraulic press (Fig. 3.15), the compaction occurs in a closed chamber. The volume of the input material is set by the chamber size and the degree of compaction, by the stroke of the piston. This principle allows for production of briquettes with a variety of shapes. The motion of the pressing piston to its maximum extension, the pressing pressure, grows exponentially based on the length of the stroke. The limiting factor is the parameters of the hydraulic circuit—maximum pressure, compaction speed, and stroke, which regulate the length of the produced briquette. However, with increased length, the compacted material loses homogeneity and thus the integrity of the pressed briquette is compromised (see Fig. 3.16). Cracks

Fig. 3.15 Briquettes produced by hydraulic press

Fig. 3.16 Defects in the structure of the briquette

become prevalent, and sufficient compaction throughout the volume of the briquette may not be obtained; therefore, not only is density affected, but more importantly the strength of the briquette is greatly reduced. This method of briquette production is based primarily on the axial pressure, which depends on the creation of shape bonds between particles within the briquette.

Pressing piston in a mechanical press—production of extrusions on mechanical presses equipped with a pressing piston occurs in an open chamber, where the input material is forced through a chamber and nozzle by means of the piston. In this method, the biomass is compacted into a complete and continuous cylindrical extrusion, which is then parted to the desired length; see Fig. 3.17. The integrity between individual pressings is achieved by a crown at the top of the pressing piston, which penetrates the end of each pressing and forms a shape, which the following pressing couples to. Since the extrusions are produced by coupling individual pressings, defects often occur at these interfaces, which reduce the mechanical strength of the extrusion; see Fig. 3.18. This principle of compaction also results in a lot of heat generated due to friction and the cyclic action of the pressing tool, which is a favorable condition when compacting biomass. At higher temperatures, the lignin within the input material plasticizes, which acts as in internal binder between particles within the extrusion. Therefore, when compared to the hydraulic pressing method, biomass briquettes produced with a mechanical press and piston tool obtain a higher density and strength.

Pressing screw of a mechanical press—briquettes made of biomass by means of a pressing screw typically obtain the highest quality of extrusion. It is possible to extrude various shapes, from circular to n-sided geometries, with center holes or

Fig. 3.17 Briquettes produces with a piston tool on a mechanical press

Fig. 3.18 Composition of the extruded briquette with visible defect

Fig. 3.19 Extrusions produced by screw-type pressing tool

Fig. 3.20 Composition and structure of extrusions—free of significant defects

without; see Figs. 3.19 and 3.20. The material is not intermittently compacted but continually, which eliminates defects at the interface of briquette sections. The extrusion is produced continuously and is parted at the desired length. The continuous extrusion ensures a high degree of compaction with good mechanical indicators of quality. The bonding of particles throughout the extrusion can be described as a combination of bonds described in both of the aforementioned methods, where various types of internal bonds are created resulting in the highest quality of extrusions. Besides all of these advantages, this method suffers from accelerated wear of the pressing screw, which is not ideal for effective production.

The process produces extrusions by means of a continuous pressing of material within a closed cycle. The input material is continuously fed through the pressing chamber and out the end of the chamber. This phase allows the extrusions to undergo a prolonged phase of acting pressure (and temperature). During this phase (also known as the calibration phase), the internal stress of extrusions is released, which would otherwise negatively affect the integrity of the extrusion as it is ejected from the chamber. Also during this phase, conditions are favorable not only for the ideal bonding of particles, but also between individual extrusions. Therefore, as the compacted material is ejected from the chamber it is theoretically possible to have an extrusion of infinite length, which in practice is parted at desired lengths. The parting of extrusion occurs either by the overhanging weight of the extrusion or mechanically with some sort of tool, i.e., knife or saw.

The aforementioned principle of compaction can be observed in mechanical BP (pressing with piston or screw tool) or in all types of pellet mills. As can be seen in Figs. 3.3 and 3.4, with plate and cylindrical dies (pressing and subsequent extrusion in an open pressing chamber).

Looking closer at the pressing process, we can see that the material is gradually forced through an array of openings in the die by means of a rotating pressing tool—rollers. This occurs with a continuous supply of input material. The material, which is pressed through the openings in the die, obtains their final shape and density. This pressing and subsequent extrusion through an opening yield a pellet, which is cut to length by means of a parting tool after extrusion through the die. This method is the same whether the die is in a disk or cylindrical configuration. The differences are in the construction of each pellet mill and the orientation of the rotating element. Disk or plate dies are usually oriented vertically as can be seen in Fig. 3.3, whereas in a cylindrical die the rotating element(s) are oriented horizontally. This difference requires a separate gearbox configuration (transmission of torque to the rotating elements).

As was discussed, pellet mills compact their material between a pair of pressing tools: rollers forcing material through an array of openings within a die by means of a rolling action. In terms of the configurations of such mills, there are some options regarding the placement of the pressing tool pair. The dies (with extrusion holes) are almost exclusively a plate or cylindrical shape. Plate (disk) dies are used in mills with vertically rotating rollers. Cylindrical dies are used in mills with vertically and horizontally rotating rollers. Increasing the back pressure and duration of the calibration phase, it is possible to obtain a change in shape of the holes in the dies. The pellet dies are manufactured with various shapes (openings or holes) cut into them

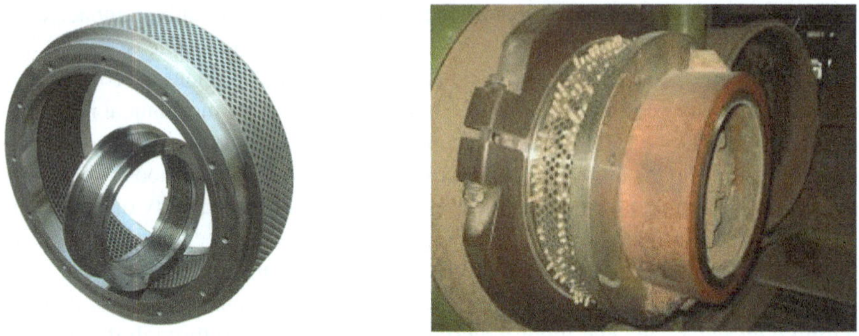

Fig. 3.21 Cylindrical pressing die in a horizontally rotating mill

Fig. 3.22 Plate (disk) die in a vertically rotating mill

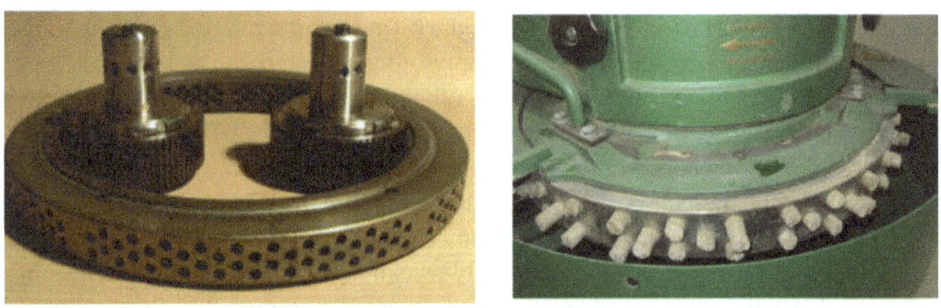

Fig. 3.23 Cylindrical die and pressing rollers (vertical rotation)

from circular to conical and even combinations of these. The following Figs. 3.21, 3.22, and 3.23 show the general configuration of the die holes.

Compacted material is pressed through the die most often by means of rollers. Rollers are made primarily with a cylindrical or conical shape. For the method to be effective, at least two rolling elements are required. There exist configurations with 3, 4, and even 5 rolling elements. Pellet mills experience the most wear in the rolling elements and pressing die. The following Figs. 3.24, 3.25, 3.26, and 3.27 show the general configuration and shapes of the rollers.

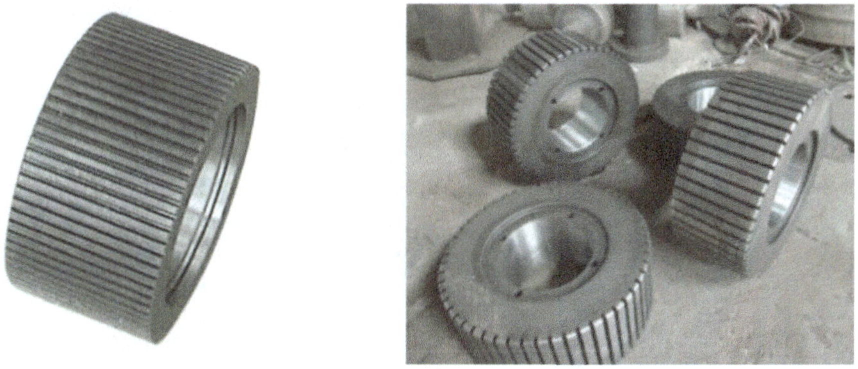

Fig. 3.24 Cylindrical rollers of a pellet mill

Fig. 3.25 Various roller configurations and orientations [14]

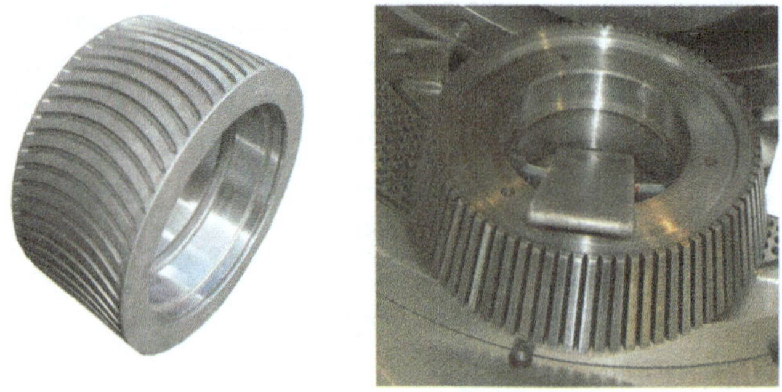

Fig. 3.26 Conical rollers use in pellet mills

Fig. 3.27 Configurations of rotating elements in mills utilizing conical rollers

The advantage of conical rollers is a more ideal distribution of pressure across the rolling surface, which results in more even wear across the surface of the pressing die. Their configurations are similar to those of cylindrical rollers.

Based on the information discussed, currently there are a variety of types, configurations, and methods used, which require a specific pressing chamber geometry. The pressing tool in compacting machines operates in extreme conditions with regard to wear and stress. Pressing machines used for the compaction of biomass, mainly wood waste biomass, experience a high level of wear compared to other types of biomass, i.e., phytomass. This is related to the friction characteristics between the material of the pressing tool (steel) and input material (wood) combined with the pressure in the chamber. For these reasons, it is necessary to specify and investigate the compaction process taking place within the pressing chamber, define the relationships between design parameters, and ultimately determine their impact on the resulting quality of extrusions. Only with extensive knowledge of the compacting process can the efficiency of the production, with respect to the type of compacted material, be improved.

3.2 The Shape Effects of the Pressing Tool

The shape of the pressing tool is tightly related to the construction principles of pressing machine and the shape of the extrusion. Analysis and optimization of the tool shape and extrusion dimension are investigated in the next chapter; therefore, the parameters are discussed later. Instead, this section investigates the shape effects of the pressing tool on the resulting quality of the extrusion.

It has been discussed that the pressing tool, in combination with the pressing chamber, is primarily responsible for the process of compaction; however, it is necessary to point out that part of the pressing chamber is a pressing nozzle; see Fig. 3.28. This nozzle is primarily responsible for the resulting shape of the briquette, since the compacted material is concentrated through the nozzle at the end of the pressing chamber. The nozzle can be of monolithic in construction or made of various parts, which can vary the internal shape (Fig. 3.29). Such variation can cause higher back pressures in the pressing chamber, which would result in a high-quality briquette; however, this higher pressure causes greater wear on the pressing tool.

Based on the type and shape of the pressing tool, as well as the construction principle, we can identify various qualities of extruded briquettes. The distribution of

Fig. 3.28 Basic parts of the pressing chamber

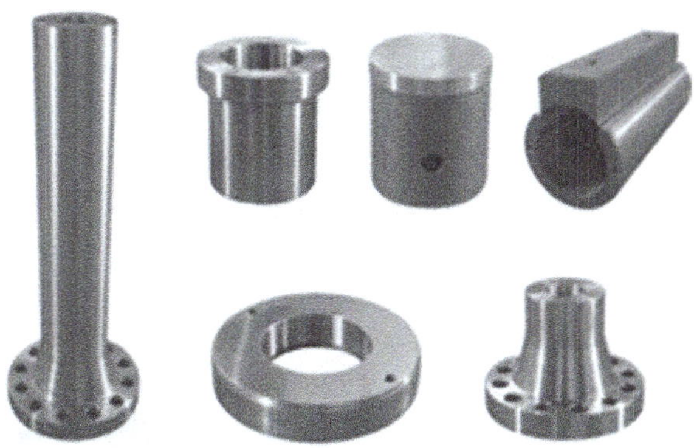

Fig. 3.29 Chamber nozzles with various shapes

forces acting along the pressing chamber resp. the pressing hole is directly dependent on the shape. As there are differences, it is important to address this issue as well. Briquettes are larger in size than pellets, and therefore, specific deficiencies in machine construction can result in defects within the extruded briquette. Internal defects significantly affect the quality of extrusions, i.e., density and mechanical integrity.

3.2.1 Surface-to-Volume Ratio

A very important parameter in the compaction process is the ratio between surface area and volume of the pressing chamber. Different shapes and sizes yield different surface-to-volume ratios. This ratio is important in the combustion process, but also affects the wear of the pressing tool. The effects on combustion are investigated within the following chapter.

Wear of the machine and its functional parts—pressing tool, is very important to know with respect to economic viability. It is therefore important to keep this in mind already when designing the pressing chamber (its size and shape) and other criteria. The greater the wear, the greater the repair costs [20]. The wear of the pressing machine also impacts the quality of extrusions. The lower the surface-to-volume ratio, the less wear occurs in the machine. This means less wear of the pressing chamber, plate/cylinder dies in pellet mills, and nozzles in briquette production.

As an example, the following Fig. 3.30 graphically compares the surface-to-volume ratio of various briquette press nozzle shapes. The resulting shape of

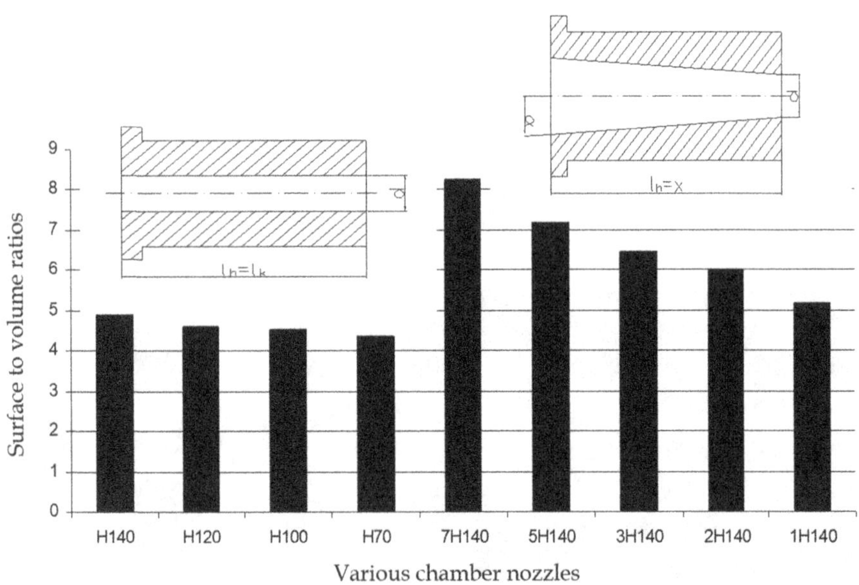

Fig. 3.30 Comparison of surface-to-volume ratios of various chamber nozzle configurations in a briquette pressing machine [21]

extrusions is directly dependent on the pressing tools; in the examples, however, the shape is the same for all cases—circular Ø 20 mm. Therefore, it is possible, by investigating the shape of the nozzle and its degree of wear expressed by the surface-to-volume ratio of the nozzle, to analyze and subsequently optimize the shape of extrusions. The nozzle marked H140, H120, H100, and H70 represent a nozzle with a cylindrical opening of Ø 20 mm, with lengths of 140, 120, 100, and 70 mm, respectively. Nozzles 7H140, 5H140, 3H140, 2H140, and 1H140 represent conical nozzle geometry with end diameter Ø 20 mm and angles 7, 5, 3, 2, and 1 degrees (over a length of 140 mm), respectively. Based on the calculated values of the surface-to-volume ratio for each of the listed nozzles, we can deduce that the least wearing part was chamber nozzle H70. Such an approach can be used to analyze various shapes of the chamber nozzles and quantify the degree of wear with respect to the shape of extrusions.

Of course, results obtained by such an analysis are necessary to supplement with data regarding the friction between materials (steel–wood, steel–straw, etc.). This means that if we know that when pressing wood sawdust, there is a higher wear of the tools than when pressing straw; we must also take this into account when designing the shape of the pressing tool. Based on the decomposition of the forces acting in the pressing chamber of the machine, we know that the friction force resp. the coefficient of bonding friction between the pressing tool and pressed material are parameters that will significantly affect the onset of failures and wear of the machine parts. Based on the results of experiments performed by German scientists [22, 23], we can state that a higher coefficient of friction will also result in an increase in necessary axial pressure acting on the nozzle (P_m) and the resulting force effects acting on the machine parts change significantly.

Parameter—shape of the pressing tool, currently has an increasing significance. Manufacturers of pressing machines are constantly trying to adapt to new trends and requirements such as the ability to change the shape of extrusions by a small adjustment or minor part change in the machine. The following Figs. 3.31, 3.32, 3.33, and 3.34 show a briquette nozzle of a crank-style briquette press, which produces a standard briquette of Ø 50 mm. By adding a pelleting insert into the briquette

Fig. 3.31 Briquette press nozzle

Fig. 3.32 Pelleting insert with four holes [24]

Fig. 3.33 Pelleting insert installed within the briquette nozzle [24]

Fig. 3.34 Comparison of extrusions—pellets/ briquette

nozzle, it is possible to change production from a Ø 50 mm briquette to a Ø 20 mm pellet using the same machine. IT is worth noting that even though the pelleting insert is of the same material as the briquetting nozzle, the friction conditions within the process are not the same. The surface-to-volume ratio is larger with the pelleting insert installed, and thus, the distribution of forces has changed, which affects the wear of the pressing tool.

The change in surface-to-volume ratio in a multichambered briquette nozzle can be investigated through parametric 3D modeling as a tool to design and optimize nozzle geometry. In the following Figs. 3.35, 3.36, 3.37, 3.38, and 3.39, the design criterion of the nozzle was to reduce diameter of openings while increasing the number of openings in the nozzle. It is obvious that some of the designs would be difficult to manufacture, but for the purposes of the demonstration they are considered.

All of the design variants above were compared in terms of their surface-to-volume ratio (Fig. 3.40), as well as their total internal volume, which can give a basis for the production capability of each design. Based on the calculated data (Table 3.1), total internal volume, and the surface-to-volume ratio, the relevant design parameters were confirmed.

The moisture content and size of input fraction of the input material are also important parameters that affect the resulting quality of extrusions and nozzle wear. Every biological material contains a certain moisture content. It is very important

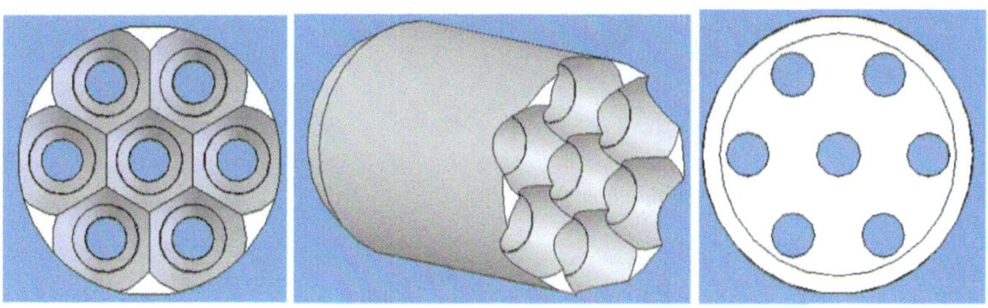

Fig. 3.35 Nozzle D10/7 (7x ø 10 mm holes) [25, 26]

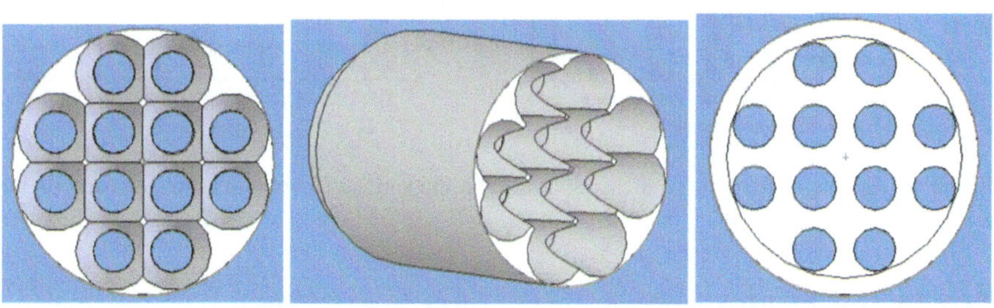

Fig. 3.36 Nozzle D10/12 (12x ø 10 mm holes) [25, 26]

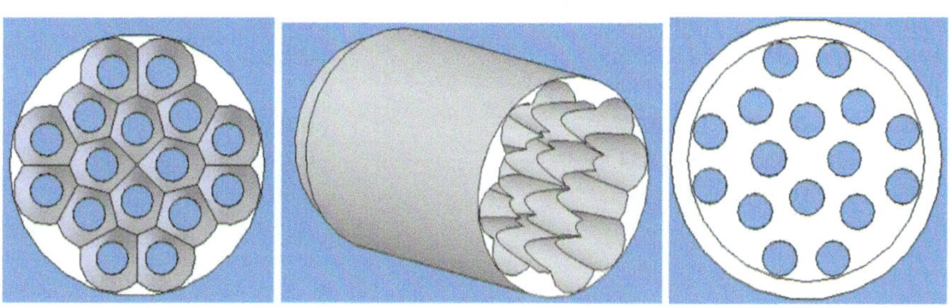

Fig 3.37 Nozzle D8/16 (16x ø 8 mm holes) [25, 26]

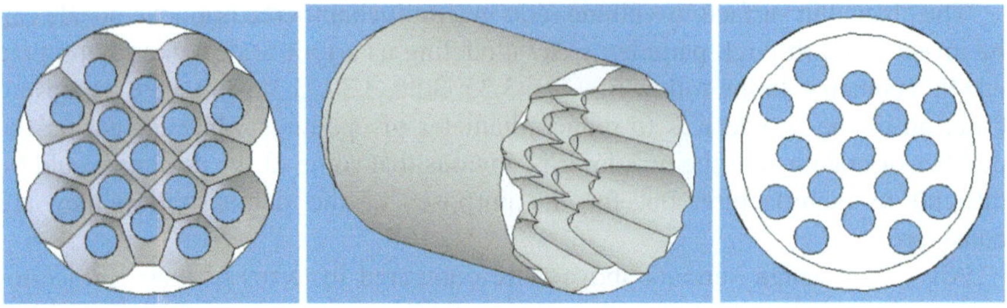

Fig. 3.38 Nozzle D8/17 (17x ø 8 mm holes) [25, 26]

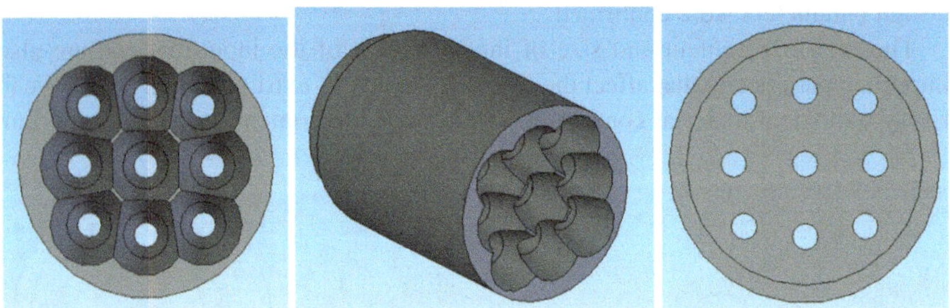

Fig. 3.39 Nozzle D6/9 (9x ø 6 mm holes) [25, 26]

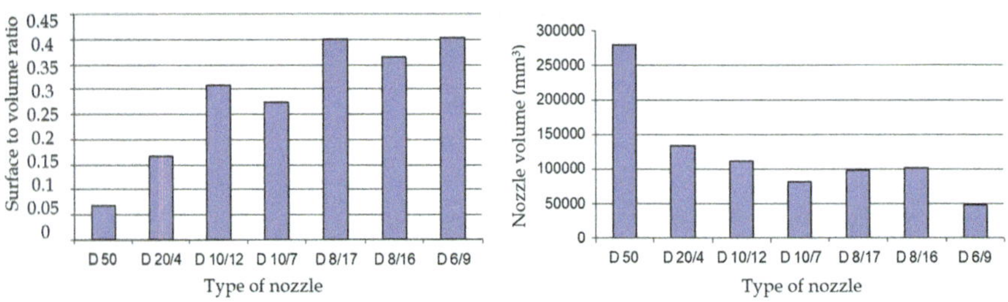

Fig. 3.40 Effects of the type of pressing nozzle on the surface-to-volume ratio (left) and on the volume of the nozzle (right) [25, 26]

Table 3.1 Calculated design parameters of each nozzle design [25, 26]

Nozzle type	Total volume—nozzle holes (mm^3)	Surface-to-total volume ratio
Briquette nozzle	279 632.4	0.0682
Pellet nozzle D20/4	134 006.6	0.1662
Pellet nozzle D10/12	111 886.5	0.3082
Pellet nozzle D10/7	81 411.1	0.2749
Pellet nozzle D8/17	97 798.7	0.4002
Pellet nozzle D8/16	102 034.8	0.3621
Pellet nozzle D6/9	47 439.5	0.4022

that this content is not too high nor too low. Based on research, the optimal moisture content of input fraction is between 10% and 16% (relative moisture) [26, 27]. If the material contains a higher content (wetter), the excess water can gather onto machine parts causing corrosion. On the other hand, if the input fraction has low moisture content (dry) the machine parts quickly accumulate heat and are subject to a higher degree of abrasive wear. Low moisture content (dry input fraction) can cause the surface of the compacted material to burn (see Figs. 3.41 and 3.42), under the pressure of the tool, and stick to the pressing machine surfaces causing accelerated wear.

It is also very important to prepare the input material fraction size, based on the material to be compacted. The input fraction size not only influences the quality of the extrusion but also the degree of wear on the pressing machine parts [28–31]. Each material has varying structure and therefore between each individual particle, a different bonding process. Of course, these bonds occur based also upon the pressing temperature and applied pressure. Comparing, for example, soft and hardwood fractions (of the same size), it is immediately evident the difference in structure. Hardwood has a "shaving"-like structure, while softwood has more of a "chip" or finer, sawdust-like structure [32, 33]. Other than the fact that each material has its specific coefficient of friction with metal, the structure of the compacted material significantly impacts the wear of the pressing machine parts. This is related to different bonding mechanisms occurring between individual particles in the compacted material, and thus the bonding force between them, but also between the material and pressing machine parts.

The acting pressure and pressing temperature are two major parameters in the process of compaction. They influence the resulting quality of extrusions to a large degree (density, abrasion and mechanical resistance, etc.). On the other hand, they also significantly impact the wear rate of the pressing machines parts since the

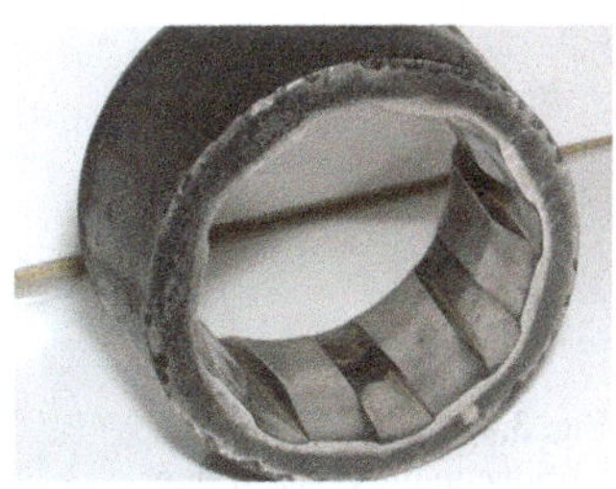

Fig. 3.41 Burnt residue of pressed material on the nozzle of a briquetting press

Fig. 3.42 Burnt residue within the groove of the nozzle

parameters are a source of loading within the entire pressing chamber. Some pressing machines allow for the setting of these parameters at certain intervals depending on the input material. Such a machine has its advantages only when the operating personnel are aware of the effects of changing these parameters mainly on the wear of the machine. However, many pressing machines have fixed parameters defined by the kinematic (mechanical) chain of the machine and the pressing temperature accumulated in the pressing chamber also cannot be controlled; however, many of these machines are equipped with a heating element, which can regulate this based on the input material.

Ignorance and nonconformity to the optimal pressing conditions result in poorly designed machinery with enormous accumulation of wear mainly due to the action of needlessly high pressing pressures; see Figs. 3.43, 3.44, 3.45, and 3.46. Tool wear is part of both briquetting (Figs. 3.41, 3.42, and 3.44) and pelleting technology (Figs. 3.43, 3.45, and 3.46). Based on the decomposition of forces within the pressing chamber, the magnitude of the pressing force directly affects the magnitude of the friction forces. This means that if we correctly balance the pressing conditions by employing optimal pressing force and temperature, we can reduce the frictional effects (Fig. 3.44) while still retaining the desired quality of extrusions. The distribution of forces within the pressing chamber can also be optimized by the shape of the pressing tool. Therefore, it is important to define and quantify the effects of

Fig. 3.43 Roller wear of a pelleting machine caused by high friction and corrosion

Fig. 3.44 Surface wear of a briquette press screw due to high frictional forces

Fig. 3.45 Wear of a disk die and pressing roller (pellet mill)

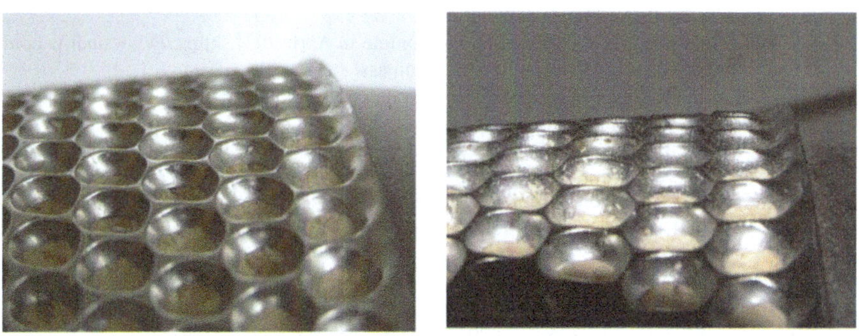

Fig. 3.46 Wear of the holes within the die of a pressing machine

construction parameters of certain pressing machine parts based on the resulting quality of extrusions.

Based on industry experience, it is possible to state that the employed method of compaction, type of pressing machine, and pressing tool directly affect the behavior and distribution of forces acting within the pressing chamber and thus significantly affect the resulting quality of extrusions. It is therefore necessary to research the effects of such parameters and the subsequent optimization of the pressing machine parts.

References

1. DIN 51731:1996, Testing of Solid Fuels – Compressed Untreated Wood, Requirements and Testing. Deutsches Institut für Normung, Berlin
2. DIN Plus: Certification scheme. In: Wood Pellets for Use in Small Furnaces. DIN CERTCO – Gesellschaft für Konformitätsbewertung mbH, Berlin (2002)
3. Ö-Norm M 7135: Compressed Wood or Compressed Bark in Natural State-Pellets and Briquettes, Requirements and Test Specifications. Österreichisches Normungsinstitut, Vienna (2000)
4. STN EN ISO 16559:2014, Tuhé biopalivá. Terminológia, definície a opis/Solid Biofuels. Terminology, Definitions and Descriptions (in Slovak)
5. Križan, P.: The Process of Pressing Wood Waste, 1. vyd, 197 p. Publishing House STU, Bratislava (2014). ISBN 978-80-227-4251-1 (in Slovak)
6. STN EN ISO 17225-2:2014, Solid Biofuels – Fuel Specifications and Classes – Part 2: Graded Wood Pellets
7. STN EN ISO 17225-3:2014, Solid Biofuels – Fuel Specifications and Classes – Part 3: Graded Wood Briquettes
8. STN EN ISO 17831-1:2015, Solid Biofuels – Determination of Mechanical Durability of Pellets and Briquettes – Part 1: Pellets
9. STN EN ISO 17831-2:2015, Solid Biofuels – Determination of Mechanical Durability of Pellets and Briquettes – Part 2: Briquettes
10. Križan, P., Svátek, M.: Research of the process of compaction of different types of biomass, 1st ed. 182 p. FX s.r.o., Bratislava (2014). ISBN 978-80-89313-59-4 (in Slovak)
11. CPM Europe BV: Company's web page, cited online in April 2021. https://www.cpmeurope.nl/en/products
12. DIN 52182 Bestimmung der Rohdichte. Deutsches Institut für Normung, Berlin
13. Amandus KAHL GmbH: Company's web page, cited online in April 2021. https://www.akahl.de/en/machines/pelleting/pellet-mill/
14. ANDRITZ AG: Company's web page, cited online in April 2021. https://www.andritz.com/products-en/group/feed-biofuel/pellet-mills-biomass
15. Kers, J., Križan, P., Letko, M.: Mechanical recycling of compounded plastic waste for material valorization by briquetting. Mater. Sci. Appl. Chem. **21**, 39–44 (2010). ISSN 1407-7353
16. Kers, J., Križan, P. Letko, M., Šooš, Ľ., Kask, Ü. Gregor, A.: Mechanical recycling of compounded polymeric waste and evaluation of briquetting parameters. In: Proceedings of the 7th International Conference of DAAAM Baltic Industrial Engineering: Tallinn, Estonia 22–24 April 2010, pp. 468–473. Tallinn University of Technology, Tallinn (2010). ISBN 978-9985-59-982-2
17. Križan, P., Matúš, M., Šooš, Ľ., Kers, J., Peetsalu, P., Kask, Ü., Menind, A.: Briquetting of municipal solid waste by different technologies in order to evaluate its quality and properties. Agron. Res. **9**(Spec.iss. 1), 115–123 (2011). ISSN 1406-894X
18. Šooš, Ľ., Kolláth, Ľ., Onderová, I., Križan, P., Ondruška, J. Čačko, V. Biath, P.: The Concept of a Separation Line for the Separation of Municipal Waste, 103 p. Bratislava STU Faculty of Mechanical Engineering (2014) (in Slovak)
19. Šooš, Ľ., Križan, P.: Variable piston of a compaction machine: Patent No. 288168 SK, 11 p. Industrial Property Office of the Slovak Republic, Banská Bystrica (2014) (in Slovak)
20. Križan, P., Matúš, M., Šooš, Ľ.: Design of pressing chamber briquetting machine with horizontal pressing axis. In: MMA 2012 – Advanced Production Technologies: Proceedings. 11th International Scientific Conference. Novi Sad, Serbia, September 20–21, 2012, pp. 361–364. Faculty of Technical Sciences, Novi Sad (2012). ISBN 978-86-7892-429-3
21. Križan, P., Šooš, Ľ., Matúš, M.: Optimalisation of briquetting machine pressing chamber geometry. In: Machine Design, pp. 19–24 (2010). ISSN 1821-1259

22. Horrighs, W.: Determining the dimensions of extrusion presses with parallel-wall die channel for the compaction and conveying of bulk solids. J. Prep Process. Aufbereitungs-Technik, nr. 12/1985, pp. 724–732, Wiesbaden (1985). ISSN 0004-783 X
23. Kegel, K.: BrikettierHandbuch, Halle (Saale), Verlag von Wilhelm Knapp (1948)
24. Šooš, Ľ., Matúš, M., Beniak, J., Križan, P.: Optimalization of the shape, dimension and density of pressing of refined fuels. In: 13th International Conference on Energy for a Clean Environment [elektronický zdroj] : Proceedings of the Abstracts. Sao Miguel, Portugal, 2–6. 7. 2017. 1. vyd. Instituto Superior Técnico, Lisboa (2017)
25. Križan, P., Šooš, Ľ., Vukelič, D.: Counter pressure effecting on compacted briquette in pressing chamber. In: MMA 2009. Flexible Technologies : Proceedings. 10th International Scientific Conference. Novi Sad, 9–10.10. 2009, pp. 136–139. Faculty of Technical Sciences, Novi Sad (2009). ISBN 978-86-7892-223-7
26. Križan, P., Šooš, Ľ., Vukelič, D.: A study of impact technological parametres on the briquetting process. Sci. J. Facta Universitatis. **6**(1), 39–47 (2009). ISSN 0354-804X
27. Križan, P., Svátek, M., Matúš, M.: Analysis of the significance of technological parameters at briquetting of selected types of hardwood and softwood. In: Aplimat 2011. Proceedings, pp. 395–404. FX s.r.o., Bratislava (2011). ISBN 978-80-89313-51-8
28. Mani, S., Tabil, L.G., Sokhansanj, S.: Effects of compressive force, particle size and moisture content on mechanical properties of biomass pellets from grasses. Biomass Bioenerg. **30**, 648–654 (2006)
29. Pietsch, W.: Agglomeration in Industry. Occurrence and Applications, p. 813. Wiley-VCH Verlag GmbH & Co. KGaA, Weinheim (2005). ISBN 3-527-30582-3
30. Pietsch, W.: Agglomeration processes. Phenomena, Technologies, Equipment, p. 614. Wiley-VCH Verlag GmbH, Weinheim (2002). ISBN 3-527-30369-3
31. Pietsch, W.: Size Enlargement by Agglomeration, p. 531. Wiley, Chichester (1991). ISBN 0 471 92991 3
32. Chuchrjanskij, P.N., Lehký, M.: Wood Finishing, Part I: Wood Pressing, p. 155. Publishing Práca, Bratislava, 1953, 674.04 (in Slovak)
33. Sitzmann, W., Buschhart, A.: Pelleting as Prerequisite for the Energetic Utilization of by-Products of the Wood-Processing Industry, 26.10.2009, Amandus Kahl GmBH & Co., Oslo

Chapter 4
Shape and Dimensional Analysis of Extrusions

The shape and size of the extrusion are mostly dependent on the design and construction of the pressing machine and the shape of the pressing tool. In the previous chapter, a theoretical analysis of the effects of pressing tool shape—pressing chamber, on the resulting quality of extrusions was discussed. The shape and size of extrusions correlate directly to the shape of the pressing chamber. Due to this fact, it is necessary to investigate the optimization of the shape and dimensions of extrusions. There are many reasons why this is important. The shape and dimensions of extrusions are directly affected by:

- Manipulation of extrusions in automated transit systems
- Storage of the extrusions
- Automated combustion processes
- Method of combustion
- Construction of the pressing tool
- Principle of compaction
- Wear of the working parts in the pressing machine
- Hourly production and operating parameters of the pressing machine

The shape and dimension of extrusions are significant due to manipulation of extrusions; this means that the shape affects the way in which the extrusions are transported, which correlates with further investment in proper transport methods and increases the investment of the entire production. Similarly, the method of feeding extrusions into the combustion chamber needs to be considered as well, which again reflects on the necessary investments (investment in combustion process, which is suitable for the given fuel) (shape of extrusions). If fossil fuels are to be replaced by biomass alternatives, it is necessary to completely automate the combustion process. Therefore, it is very important that the extrusion has a suitable shape and size [1] for trouble-free feeding to the combustion chamber by available equipment. Currently, the most common size of extrusion is pellets with a diameter Ø 6 and 8 mm. Standards dictate the maximum length to be no greater than 5 times

P. Križan, *Biomass Compaction*, https://doi.org/10.1007/978-3-030-89956-1_4

the diameter [2, 3]. The most common feed mechanism is auger or piston conveyors. STN 26 2808 defines the parameters and dimensions of auger conveyors. This standard limits the dimensions of transported material to be no greater than ¼ the diameter of the auger mechanism. Therefore, the type and size of conveyor system affect the shape of extrusions.

The shape of the extrusion is very important in terms of storage. The main parameter is to use the least amount of space for storage; however, only a few shapes meet this criterion (rectangular section, equilateral triangle section, cubes, bricks, and hexagons); however, these shapes do not allow optimal oxygenation during combustion hampering the process. In terms of the automated conveyor process, the shape and dimension of the fuel must allow users to fill them with hoses.

The shape of the extrusion has a role in the evaluation of its overall quality. The shape of the extrusion depends on the shape of the pressing tool. Wear of the pressing machines' working parts increases the maintenance costs, which are projected onto the cost of extrusions. Wear of the pressing machine parts results in lower quality of extrusions. The shape of extrusions affects the wear of the pressing machine. A general rule for extrusions is the lower the surface-to-volume ratio the better for slow burning (a desirable trait for combustion). To the same effect, pressing tool parts (arrays, nozzles, pistons, screws, rollers, etc.) will incur less wear the lower the surface-to-volume ratio. For this reason, where possible, it is desirable to find the optimum dimension of extrusions dependent on various aspects. It is therefore important to reduce the degree of wear already when proposing a shape and size of extrusion. Therefore, it can be said that the smaller the surface-to-volume ratio of extrusions, the lower the expected wear on the pressing tool of the machine while retaining the desired quality of extrusions.

4.1 Current Types of Extrusions

In the compaction process, depending on the material being pressed (wood biomass, phytomass, textiles, wood waste, agricultural waste, linen, corn, sunflower, cocoa shells, sludge from water processing, peat, paper, municipal waste, etc.), a reduction in volume of the input material can be obtained from 6 to 19 times the initial volume (cotton materials are highly compressible) [4]. Compaction is far from being a fully fledged practice due largely to the subsequent extraction of energy from extrusions. Compaction is primarily utilized in the processing of fertilizers, feed products, pharmaceutical and food products, and industrial and communal waste processing. Compaction has found its application in several areas precisely because of the reduction in volume, more efficient usability of compact units, and preservation and stabilization of certain processes. However, the diversity of compactable materials requires a specific approach to machine design and pressing tools depending on the material to be processed. Despite this, the compaction process is still widely used in the field of energy extraction. In the production of solid fuels by means of pressing machines, the applicable technical standards must be considered, which define the shape of the extrusions and the indicators of quality.

In the set of standards, STN EN ISO 17225 can observe the relationships defined in current EN standards for solid biofuels pertaining to the dimensions of briquettes and pellets; see Tables 4.1 and 4.2. In general, extrusions with diameters D < Ø 25 mm are considered to be pellets, whereas extrusions with a diameter D > Ø 25 mm are considered to be briquettes. Further, conditions for extrusions dimensions and shape are defined within the set of standards STN EN ISO 17225: solid biofuels—fuel specification and classes.

Pellets are primarily produced in a cylindrical shape; see Figs. 4.1 and 4.2. The dimensions of pellets are defined in technical standards (Table 4.1). Some of the advantages of pellets are a combustion process, which can be automated, good storage properties, relatively quick ignition, good homogeneity, etc. Some disadvantages are a lower bulk mass, which results in lower burn duration and requires modification to classic style furnaces/stoves intended for wood chops, which can be financially unfeasible.

Briquettes are produced in various shapes and sizes; see Fig. 4.3. Common shapes are cylindrical, n-sided, or brick. Cylindrical and n-sided shapes are produced with a hole. Some advantages of briquettes are their burn duration and wide application in burners/stoves/boilers without the need to modify them. They are

Table 4.1 Limit values for quality indicators for pellets based on STN EN ISO 17225-2 [2]

Indicator	Unit	Designation	ENplus-A1	ENplus-A2	EN-B
Diameter[a]	mm	D06	6 ± 1		
		D08	8 ± 1		
Length[b]	mm	L	3,15 ≤ L ≤ 40		
Mechanical durability	w-%[c]	DU	≥97,5	≥97,5	≥95,5
Bulk density	$kg.m^{-3}$	BD	≥600		
Particle size (<3, 15 mm)	w-%[c]	F	<1		

(m/m)—weight proportion;
[a]to be taken from the actual diameter of pellets;
[b]no more than max. 1% (m/m) of pellets > 40 mm;
[c] in the delivered state, % (mm)—mass proportion in percent

Table 4.2 Limit values of indicators of quality for briquettes based on STN EN ISO 17225-3 [3]

Indicator	Unit	Designation	A1	A2	B
Diameter[a] or Length[a]	mm	D	To be defined diameter, width, and length		
	mm	L_1			
Width[a]	mm	L_2	To be defined diameter, width, and length		
Height[a]	mm	L_3	To be defined diameter, width, and length		
Mechanical durability[b]	w-%[c]	DU	≥95	≥90	≥90
Particle density	$g.cm^{-3}$	DE	≥1,0	≥1,0	≥0,9

(m/m)—mass proportion
[a]Derived from actual diameter of briquette
[b]Informational test, only if traded in bulk.
[c]As delivered, % (mm)—mass proportion in percent

Fig. 4.1 Wood pellets

Fig. 4.2 Pellets made from hay

Fig. 4.3 Briquettes of different shapes and sizes

suitable for use in stoves, fireplace, boilers, etc. One of the main disadvantages is that they are unsuitable for small- or medium-sized automation.

Based on the aforementioned information, we can see that current briquette production has various shapes and sizes many of which are recognized by technical standards for biofuels. For the production of such defined shapes dictates the design of current pressing mills. However, the role of researchers is to try to find solutions that could be more beneficial. One area of interest in the field of research, design, and development of pressing machines is in the optimization of extrusion shape and size.

4.2 Theoretical Optimization of Extrusion Shape and Dimensions

Optimization of the shape and dimensions of extrusions can be achieved by modifying the shape of the pressing tool. Therefore, it is appropriate to try to theoretically analyze the shapes and dimensions of the extrusions. An important step is determining the parameters, which affect the optimization and to perform an analysis of the set parameters. As outlined above, it is possible to summarize the parameters that affect the optimization of extrusion shape and size [5–7]:

- Fuel quality—density and strength (type of compacted material and treatment technology)
- Method of combustion (differences in combustion between pellets and briquettes, porosity of extrusions)
- Machine wear (wear of the functional parts of the pressing machine and its tools)
- Automation of the combustion process (particularly when burning pellets)
- Storage and space requirements (for briquettes) and fuel delivery (pellets),
- Input/output power requirements for production

During the theoretical analysis and optimization of extrusion shape and size, it is important to determine the coefficients of the weighting factors. Since the extrusion is a fuel, the priority should be its ability to burn. There are three important parts to burning: burn duration, surface-to-volume ratio, and coefficient of porosity. Factors such as equipment wear or automation of the combustion process may have a lower priority. In this respect, storage and transport of the extrusions have the lowest priority. Criteria such as the combustion process, storage and transport, automation of the production process, and supply of the fuel to the combustion chamber are criteria according to which it is appropriate to assess the extrusion—the shape and size of the fuel. Based on practical knowledge, and from scientific research experience, the coefficients of the weighting factors can be designated as seen in Table 4.3.

4.2.1 Combustion Process of the Fuel—How the Fuel Burns

A very important parameter in the combustion process is the surface-to-volume ratio of the fuel. The lower this ratio, the longer the burn duration and therefore the duration in which we extract energy is increased. Various shapes and sizes have

Table 4.3 Coefficients of the weighting factors [4, 7, 8]

Parameter	Weighting coefficient
Combustion of the extrusion	1.3
Automatization of the combustion process	0.9
Machine wear	0.9
Storage and transport of the fuel	0.3

various surface-to-volume ratios. For analysis purposes, known geometric bodies represent the shape of the extrusions; see Table 4.4. However, other bodies are discussed to gain a broader understanding of the fuel shape effects. Two different methodologies for calculating the surface-to-volume ratio exist. The so-called unit method was calculated by software developed at our institute (Optimization V4.0). The unit volume method assumes the body to be equal to one unit (the unit volume), all but one of the dimensions of the body are then also assumed to be equal to one. The required dimension is thus expressed by the volume. Optimization V4.0 software uses input values based on formulas for the shapes, surface, and volume. The software then determines the number of variables and lists them in the operation field. The interval and step are determined for the variables, and the calculation is defined for a specific or arbitrary volume. The program substitutes all combinations of searched variables in limited steps. The variables are then written to a table, whereby comparing the values, the program searches for the maximum and minimum surface-to-volume ratio. The output shows the maximum and minimum surface-to-volume ratio, the surface and volume, and the dimensions corresponding to the maximum and minimum surface-to-volume ratio.

In Fig. 4.4, we can see the differences between values of the surface-to-volume ratio for different geometric bodies. The selected shapes of extrusions reflect shapes already produced by the industry but also some unconventional shapes. The neighboring columns for each shape represent two different methodologies in calculating the surface-to-volume ratio.

When comparing the shape of "briquettes and pellets" with respect to burn duration, the shapes fall within two extremes. If "n" extrusions have the same volume but varying surface, the object with smallest surface will burn up significantly later than an object with the largest surface. Spherical objects have the smallest surface for a fixed volume, while a flat unlimited surface with thickness approaching zero, with an unlimited number of edges "snowflake" and a core size approaching zero will burn up significantly faster. In terms of combustion, larger spherical extrusions are suitable for "slow-burning" processes such as in fireplaces, whereas fast-burning thin flakes are suitable for industrial boilers. The purpose of the analysis was to compare different geometric objects and develop them in terms of the optimal surface-to-volume ratio while fulfilling suitable storage criteria. The result of the analysis shows that, in terms of the surface-to-volume ratio, the optimal body is a sphere, while a cone shape is undesirable. Suitable objects are cylinders, elliptical cylinders, or variants of cylindrical shape. Iteration (parametric) methods (software) prove to be more suitable for such analysis in terms of mathematical methods since

Table 4.4 Shape of extrusions for analysis [4, 7, 8]

No.	Extrusion—shape	Volume—formula	Surface—formula	Figure and dimensions	
1	Sphere	$V = \frac{4}{3}\pi r^3$	$S = 4\pi r^2$	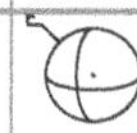	r
2	Cylinder	$V = \pi r^2 v$	$S = 2\pi r(r + v)$	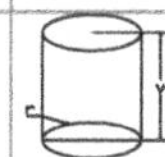	r v
3	Triangular prism	$V = \frac{a^2\sqrt{3}}{4}v$	$S = \frac{a^2\sqrt{3}}{2} + 3\cdot a\cdot v$	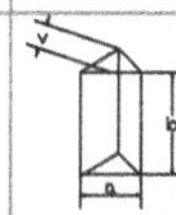	a v
4	Hexagonal prism	$v = \frac{3.\sqrt{3}.a^2.v}{2}$	$S = 3.\sqrt{3}.a^3 + 6.a.v$	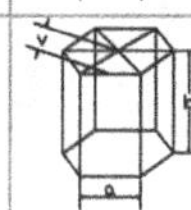	a b V
5	Cone	$V = \frac{1}{3}r^2 v$	$S = \pi r(r+s)$ kde $s = \sqrt{v^2 + r^2}$	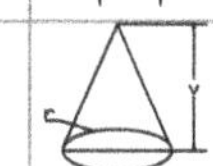	r v
6	Cylinder with hole	$V = \pi v(R^2 - r^2)$	$S = 2\pi(R^2 - r^2 + Rv + rv)$	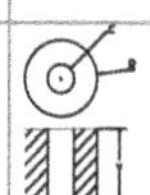	r R v
7	Cube	$V = a^3$	$S = 6a^2$	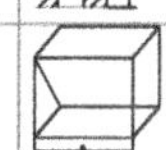	a
8	Elliptical cylinder	$V = \sqrt{a^2}\cdot\sqrt{b^2}\cdot\pi\cdot v$	$S = 2\cdot\pi\cdot a\cdot b +$ $\pi\cdot\left(\frac{3}{2}(a+b)\cdot\sqrt{a\cdot b}\right).v$	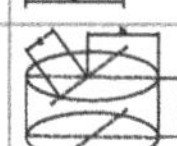	a b v
9	Intersecting cylinders	$V = 2,323r^3$	$S = 10,752r^2$		r

they converge on solutions based on set parameters such as shape and volume. Utilizing numerical solutions, we determined not only comparisons between the surface and volume of certain objects, but also the shape ratio for a unit volume and surface. The solution was performed in 0.01 increments within intervals ranging from 0.1 to 10 for dimensions describing the object.

For fast burning of the extrusion, the surface-to-volume ratio must be as high as possible, whereas for slow-burning processes it should be as low as possible. Therefore, the most suitable shapes are either spherical or quasi-spherical with sharp edged, depending on the desired effect. The surface roughness, in the case of objects with sharp edges, contributes to quick burning, whereas a low surface-to-volume ratio allows for longer more even burning. New shapes were also determined in the form of a two intersecting cylinders. Combustion of briquettes and pellets in residential boilers undergoes a pyrolytic reaction [10]. This is based on the

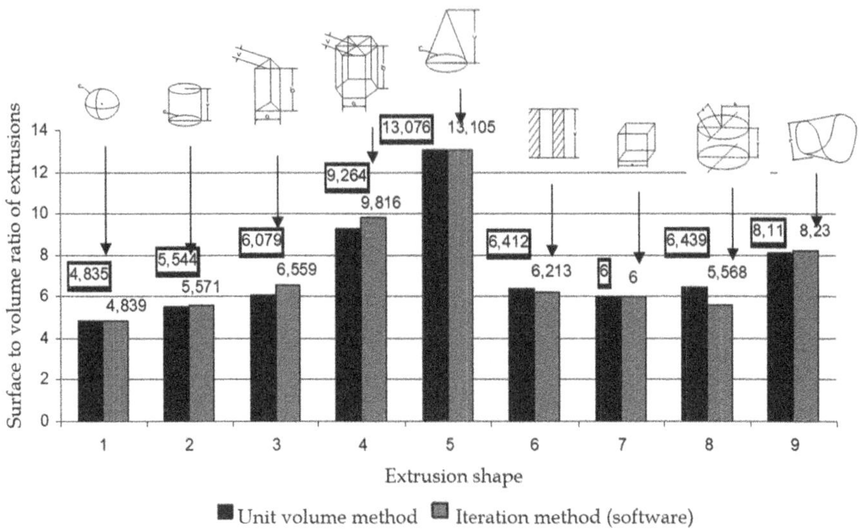

Fig. 4.4 Comparison of surface-to-volume ratios of different extrusion: *1*—sphere, *2*—cylinder, *3*—triangular prism, *4*—hexagonal prism, *5*—cone, *6*—cylinder with hole, *7*—cube, *8*—elliptical cylinder, *9*—intersection of two cylinders [9, 10]

fact that other than biomass no other raw material behaves ecologically when burned. Briquettes and pellets have the same shape, moisture, and density when burned. Compacted waste consistently has better mechanical properties than the base materials alone (density of extrusion from wood biomass is ρ = 1000 to 1400 kg.m^{-3}, whereas a common spruce tree exhibits a density of around ρ = 400 kg.m^{-3} [11, 12]).

The briquettes and pellets burn continuously, with a steady flame, due to the almost homogeneous composition of the material (Figs. 4.5 and 4.6) [13, 14]. Burning time depends on the shape, size, and type of material. The published analysis results have shown that the shape and size of the briquette diameter have an effect on the burning rate. Furthermore, the results of laboratory tests have shown that briquettes and pellets of square and rectangular section have, in principle, a greater loss of moisture, which causes a faster onset of burning at the corners and edges, whereas cylindrical briquettes burn slower. Larger diameter briquettes tend to burn more slowly, thus supplying heat for longer.

Pellets and briquettes can be optimized in terms of their parameters. Important parameters are the burn duration, the surface-to-volume ratio of the extrusion, and the size of and gaps for suitable air ingress during combustion. The availability of oxygen is important during combustion, and therefore, the interstitial of the extrusions is important (Fig. 4.6). This parameter is discussed in more detail throughout the next part of this chapter. A spherical shape appears to be the optimal shape in terms of openings within extrusions. The space factor of a sphere is 0.427, whereas the space factor for a cube is equal to zero. This implies that air ingress for a cube is zero. The entire burn duration for briquettes and pellets consists of the heating,

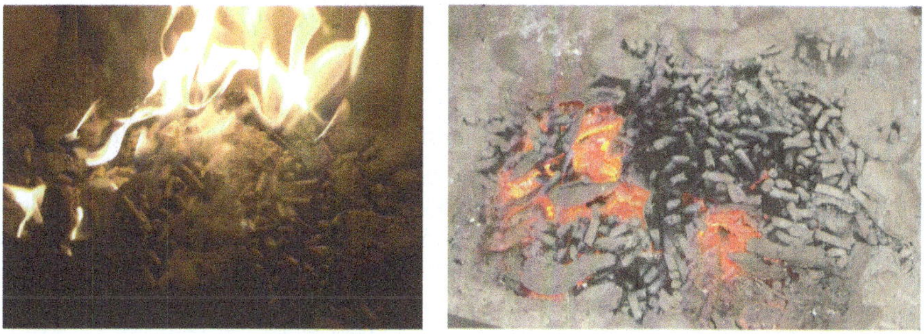

Fig. 4.5 Pellet combustion (left), pellets burning (right)

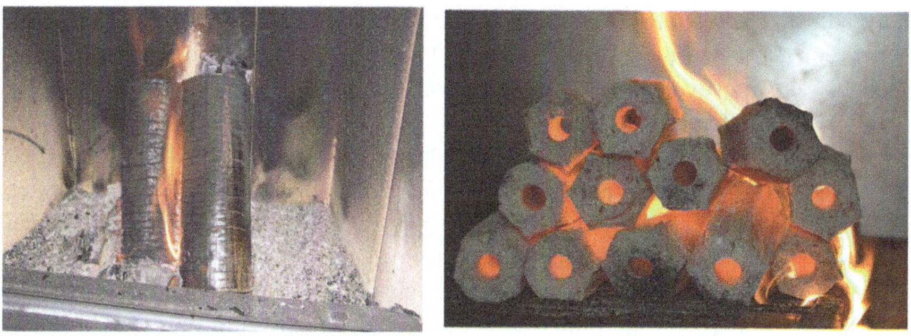

Fig. 4.6 Combustion of briquettes (left), importance of gaps and edges in the extrusion during combustion (right)

ignition, water evaporation, and degassing time. The total duration cannot be taken simply as the sum of these individual times. For example, the burning time of solid combustibles begins already as the volatile combustibles are consumed. To determine the burn duration of briquettes and pellets, we can approximate it through the use of the empirical formula for the burning of coal fuel, the so-called Nusselt's criterion (see Eq. 4.1) [8, 15]. This criterion takes into account the conditions of the combustion chamber and the size of the object.

$$\tau_h = \frac{\rho_n . g . 273 . V_{V\,min} . \alpha_v . r_Z^{\ 2}}{2 . D_0 . T_m} \quad (4.1)$$

Where ρ_n—fuel density (kg. dm^{-3})
g—gravitational constant (m. s^{-2})
V_{Vmin}—volumetric consumption of air (m^3. kg^{-1})
α_v—residual air factor
r_z—characteristic dimension of the extrusion—diameter (mm)
D_0—diffusion coefficient at 0 °C (m^2. h^{-1})
T_m—the core temperature of the flame (°K)

4.2.2 *Automation of the Combustion Process of Extrusions*

In analyzing the shape and dimensions of the extrusions, when optimizing the shape, it is necessary to consider the automation of the process. We are mainly focused on the automation of the combustion process in residential homes.

It is known that briquettes are consumed in home boilers, or fireplaces in boilers from 10 to 100 kW with characteristic configuration combustion or gasification processes. In gasification process, the chamber is filled with fuel (briquettes), which usually lasts from 4 to 10 (Fig. 4.7). Currently, it is not viable to automate these processes. Wholesalers, on the other hand, provide chambers, which are capable of integrating the briquettes into an automatic combustion process, from the fuel container to the combustion chamber. Fuel dosing into the combustion chamber is suitably adjusted according to the type of fuel burned and can be solved by using a screw conveyor or by supplying an even layer of fuel by a belt conveyor. Grate feeders, rake conveyors, and turnstiles are also often used for fuel supply.

Due to their shape and dimensions, pellets are mainly suitable for combustion in smaller pellet boilers used in residential houses or smaller buildings. For these applications, these devices are now mostly supplied as a complex automated system consisting of a pellet hopper, automated dosing and conveying systems, and the pellet boiler itself. Many global manufacturers prefer pellet dispensers in the range from ø 4 to ø 25 mm. Sizing automated combustion, transport, and dosing systems also largely depend on the dimensions of the fuels—pellets to be used (Fig. 4.8). Otherwise, problems will occur. Pellets of larger (unsuitable) dimensions for conveyors (usually for pellets longer than 45 mm) can cause blockages of the conveyor systems and problems in the control and monitoring of the pellet combustion

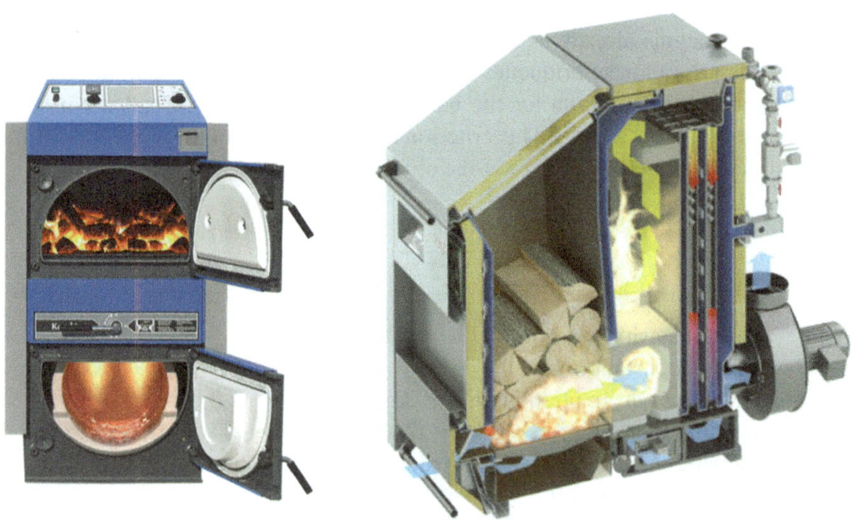

Fig. 4.7 Combustion equipment—wood and briquette gasification heaters [9]

Fig. 4.8 Part of the automated combustion process where problems may occur due to unsuitable size of input material (fuel) [6]

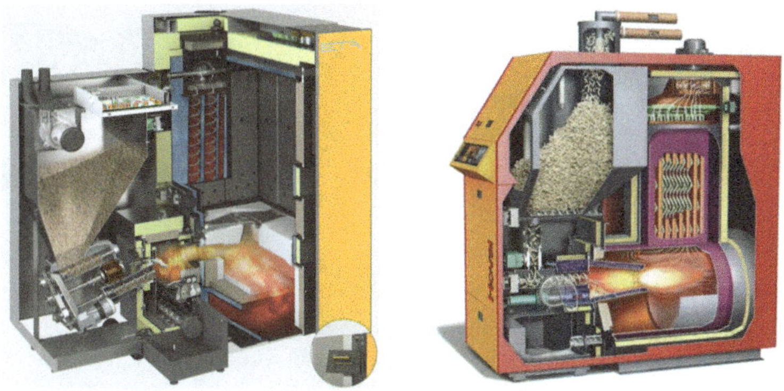

Fig. 4.9 Pellet burning equipment with integrated hopper [9]

process. The STN ISO 1050 standard specifies parameters and main dimensions for screw conveyors. The standard stipulates that the largest dimension of the transported material must not exceed 1/4 of the screw diameter (for screw feed).

There are several ways of dosing fuel resp. concepts of combustion machines with automated pellet dosing system (Figs. 4.9 and 4.10). These are commonly screw conveyors and feeders (domestic boiler rooms) and pneumatic transportation. For screw pellets in low output boilers, only screw conveyors are used, mainly due to the simple control of the feed screw. The fuel should also be optimized with regard to the dosing systems and using dispensers capable of handling the prescribed fuel dimensions. For large combustion systems (boiler rooms), the use of pellets is inefficient and not recommended. The combustion of fuel takes place on a grate, while the filling process is fully automated by means of screw conveyors, piston dispensers, or grate feeders and even combinations of these systems (Fig. 4.9). These devices are designed for larger fuel sizes. Similar conditions apply for the use of pneumatic transport and supply systems. It is inadmissible for the fuel supply to be blocked during transport of the pellets due to incorrect dimensioning. All these regularities must be considered when optimizing the shape and dimensions of the extrusions that the producer intends to create.

Fig. 4.10 Supply and dosing methods of pellets [5, 9]

4.2.3 Wear of the Functional Parts of the Device—The Pressing Tools

When designing the extrusion (its size and shape), the surface-to-volume ratio of the extrusion is an important indicator that directly affects the wear of the machine and its functional parts—the pressing tools. Generally speaking, the lower the surface-to-volume ratio of the extrusions, the less wear occurring in the pressing tools, which in turn affects the quality of extrusions. Several biofuel producers can accurately estimate the wear of the pressing tools based on the quality of the biofuels produced. The significance and importance of this ratio, especially from the point of view of the wear of pressing tools, were analyzed in the previous Chap. 3.

4.2.4 Storage and Transport

Storage, transport, and handling depend on the shape of the extrusions. For each specific shape, it is necessary to design a method of storage and transport. Storage, transportation, the combustion process, and automation of it are dependent on the storage and transport. The approximation of the problem is as follows: An equilateral triangle and a square are planar shapes that fill a 2D space such that no spaces result (see Fig. 4.11). A regular hexagon with the same size edges, whose adjacent edges form an angle of 120° and three pairs of edges together add up to 360°, is more suitable. A hexagon is also a shape that completely fills a 2D space.

How is this applicable in a 3D space? Firstly, it is necessary to take into account the manufacturability of the extrusion. It is most advantageous to optimize the shape of the extrusions so that it has a cross section of a certain planar shape "extruded" into space. Extrusions produced in this way hold their shape and have a high degree of compaction, an almost arbitrary profile and high quality. The second option is to use compaction, whereby the final product is a granule of the desired shape. Filling a space is convenient for storage, but not necessarily for burning. During combustion, it is necessary to provide a flow of oxygen. In terms of airflow and optimum gap, a spherical shape should provide optimum spacing. Effective storage, packaging, transport, and handling of extrusions are influenced by bulk permeability, which is in turn influenced by [4, 9, 10]:

- *Particle shape*—for example, stacking cubes would result in a bulk permeability of $\varepsilon = 0$, whereas storing spherical objects would never result in $\varepsilon = 0$.
- *Fractional compositions*—extrusions are mono-disperse, the same size, and shape. Bulk permeability is larger than poly-disperse substances where pores can become filled with dust.
- *Particle arrangement*—spherical particles of identical diameter have 6 ways of arrangement and 6 different values of ε. The permeability also depends on the number of contacts between the particles. The permeability is highest at the container wall.
- *Proximity to the container wall*—the space can be calculated according to the following Eq. 4.2, where V_ε [mm^3] is the volume of the space and V_t [mm^3] is the volume of the extrusion:

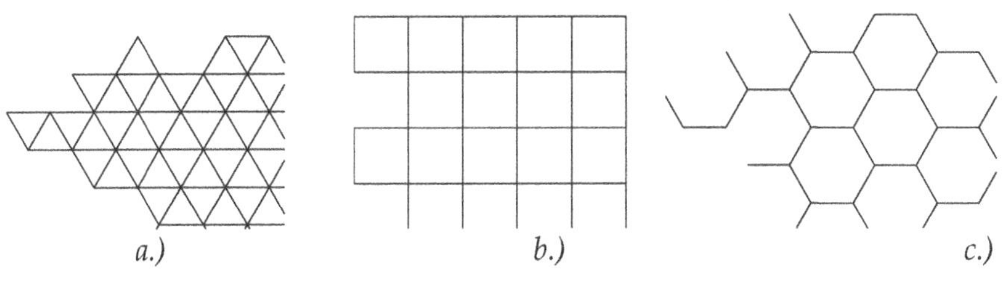

Fig. 4.11 Filling a 2D space with planar shapes: (**a**) triangles; (**b**) squares; (**c**) hexagon [10]

$$\varepsilon = \frac{V_\varepsilon}{V_\varepsilon + V_t} \tag{4.2}$$

If the storage consists of particles of the same geometric shape and arrangement, then the free space does not depend on the size of the object (see Fig. 4.12). However, if we compare different shapes of extrusions, i.e., different geometric shapes with consideration of the same arrangement (Fig. 4.13), the size of the free space changes. This parameter subsequently affects several operating costs. The practical benefit of considering the bulk permeability in briquettes is mainly in storage and transport, illustrated in Fig. 4.14. The output of the production is briquettes stored on a pallet and wrapped in foil or wrapped in paper bags. Bulk permeability is almost impossible to determine for freely stored objects, e.g., in a container or in a sack. It is very difficult and demanding to determine the arbitrary orientation of extrusions within a space, despite the fact that nowadays there are several useful software tools such as EDEM. Also for this reason, it is more important to consider this parameter only for briquettes. For pellets, it is necessary to simplify this task and place the objects in the calculation within a block, the dimensions of which determine the maximum dimensions of the optimized extrusion. Even a slight change in the size of the extrusions (also a change in their shape) can significantly affect the bulk permeability.

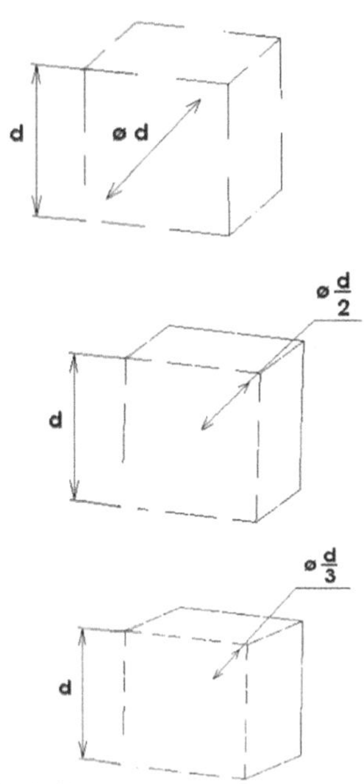

$$V_\varepsilon = V_{\mathrm{kocky}} - V_{gule}$$

$$V_\varepsilon = d^3 - \frac{\pi . d^3}{6} = d^3\left(1 - \frac{\pi}{6}\right) \tag{4.3}$$

Bulk permeability

$$\varepsilon = \frac{V_\varepsilon}{V_{kocky}} = \left(1 - \frac{\pi}{6}\right) = 0{,}427 \tag{4.4}$$

$$V_\varepsilon = d^3 - 2^3 \frac{\pi}{6} \cdot \left(\frac{d}{2}\right)^3 = d^3\left(1 - \frac{\pi}{6}\right) \tag{4.5}$$

$$\varepsilon = \left(1 - \frac{\pi}{6}\right) = 0{,}427$$

$$V_\varepsilon = d^3 - 3^3 \frac{\pi}{6} \cdot \left(\frac{d}{3}\right)^3 = d^3\left(1 - \frac{\pi}{6}\right) \tag{4.6}$$

$$\varepsilon = \left(1 - \frac{\pi}{6}\right) = 0{,}427$$

Fig. 4.12 Arrangement and calculation of bulk permeability in spherical particles of different diameters within the same bulk volume [4, 9]

Fig. 4.13 Arrangement and bulk storage of different briquette shapes

Fig. 4.14 Arrangement of briquettes to a uniform size—storage and transport

One of the basic criteria for optimizing the shape and size of extrusions should be maximizing the storage space even in non-oriented storage.

At the output of a pellet production line, pellets are mostly packed in plastic bags weighing 15 kg. Packing pellets in bags weighing more than 20 kg is not suitable for residential buildings, because such bags are very heavy for household use and are difficult to handle. For more industrial applications, it is possible to pack the pellets into so-called big bags (textile bags) weighing a maximum of 1000 kg. However, pellets can also be transported to the customer in tanks. Tanks on trucks are filled with pellets directly in the production line, which is subsequently delivered to the consumer on request. The consumer's container is filled using a pneumatic system, which allows a fully automatic supply method.

Based on the above information, it can be seen that the shape of the extrusions and the shape of the pressing tools directly affect the various aspects that are part of extracting energy from biomass. For this reason, it is important and justified to take into account all processes related to extrusions when optimizing the design of pressing tools. The solution must be approached globally and thoroughly analyze the effects of modifying the shape of the pressing tool based on the economics of the energy extraction process. On the other hand, by optimizing the shape and size of the extrusions, it is possible to achieve the optimal shape of the pressing tool for the desired output. The aim of this monograph is to provide a comprehensive picture of the issues addressed herein.

References

1. STN EN ISO 17829:2015 Solid Biofuels – Determination of Length and Diameter of Pellets
2. STN EN ISO 17225-2:2014, Solid Biofuels – Fuel Specifications and Classes – Part 2: Graded Wood Pellets
3. STN EN ISO 17225-3:2014, Solid Biofuels – Fuel Specifications and Classes – Part 3: Graded Wood Briquettes
4. Šooš, Ľ.: Graded energy carriers based on new types of raw materials. In: 6. International Slovak Biomass Forum – ISBF 2007: Proceedings. Bratislava, 19.2. 20.2. 2007. Energy Centre, Bratislava (2007). ISBN 978-80-969646-0-4 (in Slovak)
5. Križan, P., Svátek, M.: Research of the process of compaction of different types of biomass, 1st ed. 182 p. FX s.r.o., Bratislava (2014). ISBN 978-80-89313-59-4 (in Slovak)
6. Križan, P.: The Process of Pressing Wood Waste, 1. vyd, 197 p. Publishing House STU, Bratislava (2014). ISBN 978-80-227-4251-1 (in Slovak)
7. Šooš, Ľ., Križan, P., Matúš, M.: Solid graded fuels as an alternative to conventional energy carriers. In: Energetika a biomasa seminar, pp. 62–72. ČVUT, Praha (2010) (in Slovak)
8. Šooš, Ľ., Matúš, M., Beniak, J., Križan, P.: Optimalization of the shape, dimension and density of pressing of refined fuels. In: 13th International Conference on Energy for a Clean Environment [elektronický zdroj] : Proceedings of the Abstracts. Sao Miguel, Portugal, 2–6. 7. 2017. 1. vyd. Instituto Superior Técnico, Lisboa (2017)
9. Šooš, Ľ., Matúš, M., Križan, P.: Standardization of solid graded biofuels. In: Power Engineering 2010. Renewable Energy Sources: 1st. International Scientific Conference OZE. Tatranské Matliare, SR, 18–20. 5. 2010. STU in Bratislava, Bratislava (2010). ISBN 978-80-89402-24-3 (in Slovak)
10. Šooš, Ľ., Kolláth, Ľ.: Optimization of the shape, size and density of compacted fuel extrusions. In: Vykurovanie 2017: Proceedings of 25th International Scientific Conference. Podbanské, SR, 6. – 10. 3. 2017. 1st. ed. pp. 243–249. SSTP, Bratislava (2017). ISBN 978-80-89878-06-2 (in Slovak)
11. Chovanec, D., et al.: Wood Science I., Textbooks, p. 277. University of Forestry and Wood Technology in Zvolen (1985), 674.03(075.8) (in Slovak)
12. Požgaj, A., et al.: Structure and Properties of Wood, 485 p. Publihsing House Príroda a.s, Bratislava (1997). ISBN 80-07-00960-4, (in Slovak)
13. Basu, P.: Biomass Gasification and Pyrolysis: Practical Design and Theory, p. 365. Elsevier Inc, Amsterdam (2010). ISBN 978-0-12-374988-8
14. Chuchrjanskij, P.N., Lehký, M.: Wood Finishing, Part I: Wood Pressing, p. 155. Publishing Práca, Bratislava, 1953, 674.04 (in Slovak)
15. Šooš, Ľ., Križan, P., Beniak, J., Matúš, M.: New designs of compacting machines and shapes of pressing bio-biofuels. Int. Sci. Index. **17**(10, part IV), 550–555 (2015). ISSN 1307-6892

Chapter 5
The Effects of the Pressing Chamber Design Parameters on the Quality of Extrusions

Each production facility must ensure a certain quality of production. The quality of briquettes and pellets from biomass as an alternative biofuel is set by standards. These standards indicate the geometric dimensions of the pellets, as well as their density, moisture, bulk density, mechanical resistance, calorific value, ash content, and the content of chemical elements in the extrusions. Indicators of extrusion quality, in terms of the compaction process, are the mechanical properties (strength, density, mechanical resistance) [1], which depend on many factors. In order to achieve the quality specified by the standard, it is necessary to understand the influence of each individual parameter on the quality of extrusions. In the field of biomass compaction, research is mostly involved with gaining knowledge regarding the behavior of important parameters, their impact on various aspects and factors of the biomass compaction process, and its effect on the compaction process. Based on experience and detailed analysis on the impact of parameters on the biomass compaction process, it is important to pay attention to the properties of these parameters in experiments (see below). The analyzed parameters worthy of experimental research for their influence must be divided into three groups since they are not of the same character. The first group of parameters is related to the type of material used for the experiments. The second group of parameters is related to the compaction process itself, i.e., the pressing technology. The third group deals with the design parameters of the pressing machine.

A verifiable assumption can be made regarding the quality of the extrusions being influenced not only by technological parameters such as the pressing pressure, pressing temperature, and material parameters (moisture of the input material or fraction size), but also by design parameters of the pressing machine itself [2–4]. The pressing machine design parameters are [5, 6]:

- Diameter of the pressing chamber
- Length of the pressing chamber
- Taper of pressing chamber walls

P. Križan, *Biomass Compaction*, https://doi.org/10.1007/978-3-030-89956-1_5

- Shape, type, and geometry of the press tool
- Pressing method
- Friction parameters of the pressing process (binding friction coefficient)
- Length of cooling channel
- Material properties of the pressing chamber and the pressing tool

In terms of the pressing machine design, all of the abovementioned parameters are valuable, but based on the possibilities of experimental machinery and due to correct evaluation of results, the complexity of the whole process, and accurate interpretation of results, it is necessary to divide the experiment into several stages [7, 8]. It is obvious that a detailed investigation of the effects and impact of these parameters requires extensive and thorough analysis of the issue. For this reason, it is necessary to develop a theoretical analysis of the impact of these design parameters, which was the basis for the design of the experimental research. In this analysis, knowledge gained from foreign research institutes and from experience in practice was used.

5.1 Effects of Pressing Chamber Length

Currently, developers of pressing mills have not considered the pressing chamber length to be of much importance. However, based on past theoretical analyses and experiments, the length of the pressing chamber has a significant effect on the final quality of extrusions. The length of the pressing chamber significantly contributes to the distribution of pressure within the pressing chamber during uniaxial compaction on both vertical and horizontal presses. The length of the pressing chamber affects the resulting strength and density of extrusions. If the chamber is too short, the extrusion will not reach sufficient strength and the residual radial stresses will cause rupture immediately after leaving the pressing chamber. If the chamber is too long, there can be an enormous increase in back pressure created by the length of the pressed column, causing blockage of the machine and consequent damage. It is therefore important to find the optimal relationship between pressing pressure and the length of the pressing chamber with regard to the resulting quality of extrusions, resp. find a relationship determining the length of the pressing chamber as a function of the pressing pressure, or the length of the pressing chamber, which will prevent rupture of the extrusion after exiting the chamber.

To theoretically verify the effect of the pressure chamber length, it is necessary to use a certain mathematical model, which describes the effect of the pressures within the pressing chamber with respect to the length of the pressing chamber. As a result of a thorough analysis of the available scientific literature, there is some knowledge of existing mathematical models that can be used for this purpose.

The authors [9, 10] present a mathematical model describing the pressure conditions in a closed pressing chamber during uniaxial pressing on a vertical press (Fig. 5.1), where the back pressure is induced by a plug at the end of the chamber.

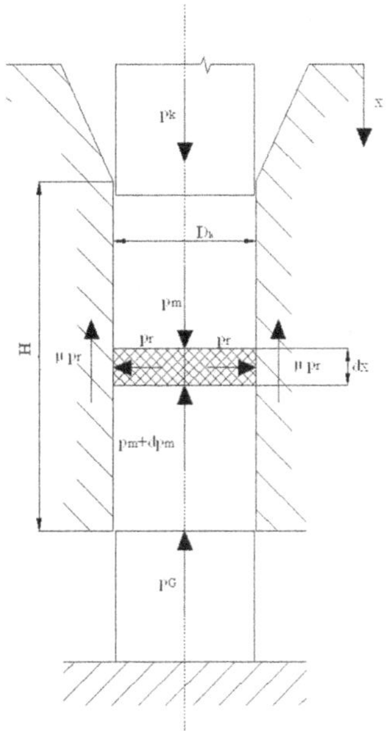

p_k – axial pressure on piston (MPa)
p_G – back pressure (MPa)
p_r – radial pressure (MPa)
p_m – axial pressure on plug (MPa)
D_k – diameter of pressing chamber (mm)
μ – coefficient of friction (-)
H – length of pressed plug (mm)

Fig. 5.1 Pressure ratios in the chamber during uniaxial vertical compaction [9, 10]

The pressure conditions in the pressing chamber between the piston and the plug are explained and described by the author in Refs. [9, 10] on an element of pressed material dx, neglecting the weight of the extrusion throughout the pressing process.

From the pressure conditions in the chamber, it is possible to write an equation for the balance of forces:

$$\left[p_m - \left(p_m + dp_m\right)\right]\frac{\pi D_k^2}{4} - \mu p_r \pi D_k dx = 0. \tag{5.1}$$

When pressing bulk materials, pressure anisotropy is usually considered (higher pressures act perpendicularly). The ratio of the principal stresses (radial σ_r/axial σ_m) is referred to as the residual pressure coefficient or the horizontal compression ratio λ. For dispersive materials, $\lambda \in \langle 0 \div 1 \rangle$.

$$\lambda = \frac{\sigma_r}{\sigma_m} = \frac{p_r}{p_m} \Rightarrow p_r = \lambda p_m \ \left(\text{MPa}\right) \tag{5.2}$$

After substituting (5.1) into (5.2) and solving for the boundary conditions: $x = 0$, $p_m = p_k$, for $x = H$, $p_m = p_G$, based on Ref. [9, 10], it is possible to obtain the following relation:

$$p_k = p_G.e^{\frac{4.\lambda.\mu.H}{D_k}} \ (\text{MPa}). \tag{5.3}$$

Equation (5.3) gives the relationship between the piston pressure p_k and the back pressure acting on the extruded material p_G. This relationship can be applied to open chamber pressing (Fig. 5.2), where the back pressure is created by a static frictional resistance between the pressed material and chamber walls. This relationship can be further used to determine the amount of back pressure required for the pressing operation and to determine the static frictional resistance of the already extruded column. Experience has shown that, after pressing, a certain amount of pressure is still required to overcome the compacted "plug."

In Ref. [9, 10], the authors also focus on the formulation of a relationship for the compacted plugs, since after releasing the load acting on them, residual radial pressure persists. This causes the extrusion to rupture. Therefore, it is very important to find the relationship between the pressing pressure and the length of the chamber (calibration part), respectively find a relationship that determines the length of the pressing chamber as a function of the pressing pressure, the length that does not cause rupture of extrusions after leaving the pressing chamber. During compaction, the primary stress is acting axially. However, after the pressed material exits the chamber the residual stresses act radially. This change in direction of the primary stress must occur during the unloaded phase—as the extrusion exits the chamber. Therefore, there must exist a state where the radial and axial stresses are equal (hydrostatic pressure state). After reaching this state, the radial stresses overcome the axial stresses until only the residual radial stress $\sigma_{r,r}$ remains. The pressed material must be extruded from the pressing chamber only in the phase where the strength of the extrusion is greater than the residual radial stress $\sigma_{r,r}$, resp. $p_{r,r}$. If this condition is met, the pressed material can be extruded through the calibration part of the pressing chamber.

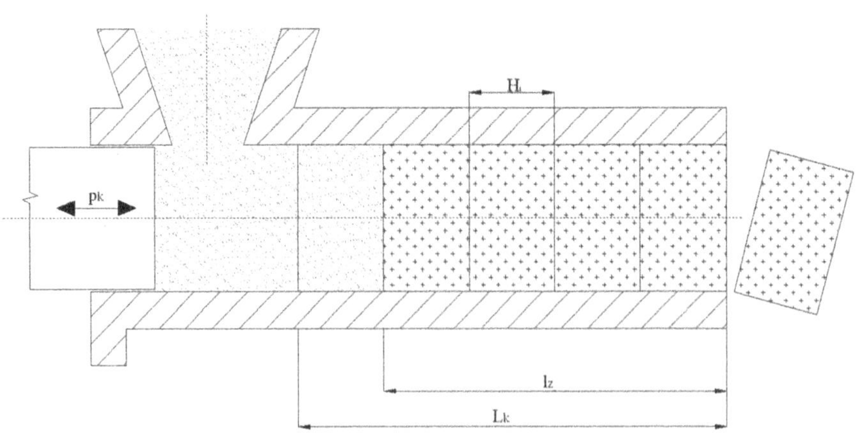

Fig. 5.2 Pressing chamber with individual extrusion "plugs" [11]

This model was used to mathematically investigate the influence of pressure chamber length on the resulting quality of extrusion. While investigating the effect of changing pressing chamber length, a length of 140 mm (the maximum allowable length of the experimental pressing chamber) was chosen as an initial value (see next chapter). This value was decreased by 10%. The remaining parameters were chosen according to the experimental apparatus. The behavior of the axial pressure on the plug p_m was monitored due to the change in length of the chamber, since the axial pressure on the plug dictates the resulting strength of the extrusion. To express the axial pressure on the plug, it was necessary to estimate the following proportions of applied pressures:

$$\lambda = \frac{\sigma_r}{\sigma_m} = \frac{P_r}{P_m} \in 0 \div 1 \Rightarrow \text{selected} \rightarrow 0{,}5 \quad (5.4)$$

$$\alpha = \frac{P_r}{P_k} \cong 0{,}15 \div 0{,}20 \Rightarrow \text{selected} \rightarrow 0{,}175. \quad (5.5)$$

Then $\frac{P_r}{P_k} = 0{,}175 \Rightarrow P_r = 0{,}175.P_k$ substituting (5.5) into (5.4), we get Eq. (5.6)

$$P_k = 2{,}86.P_m \quad (5.6)$$

Solving for Eq. (5.3), we obtain the relation (5.7), which is used in the mathematical analysis

$$P_m = \frac{P_G}{2{,}86}.e^{\frac{4.\lambda.\mu.L}{d}} \quad (5.7)$$

It is possible to see the calculated data in Table 5.1. In Fig. 5.3, these data are expressed graphically. Based on the above, it can be stated that the length of the pressing channel (chamber) has an effect on the character of the pressures within the pressing chamber and thus on the resulting quality of extrusions. Since this is a theoretical analysis and since the mathematical model used does not consider the influence of the parameters on the density of the extrusions, it is only possible to show graphically the dependence of axial pressure on the plug with respect to the change in pressing chamber length. However, based on practical experience and further experimental research at our institute, the dependence of extrusion density on the pressing pressure is also available (see Fig. 5.4). Based on the above

Table 5.1 Calculated value of pressing chamber length [12, 13]

L (mm)	**140**	126	113.4	102	91.8	82
Δ L (%)	–	−10	−10	−10	−10	−10
Pm (MPa)	**46.95**	28.77	18.51	12.42	8.7	6.17
Δ Pm (%)	–	−38,7	−35,7	−32,9	−29,95	−29

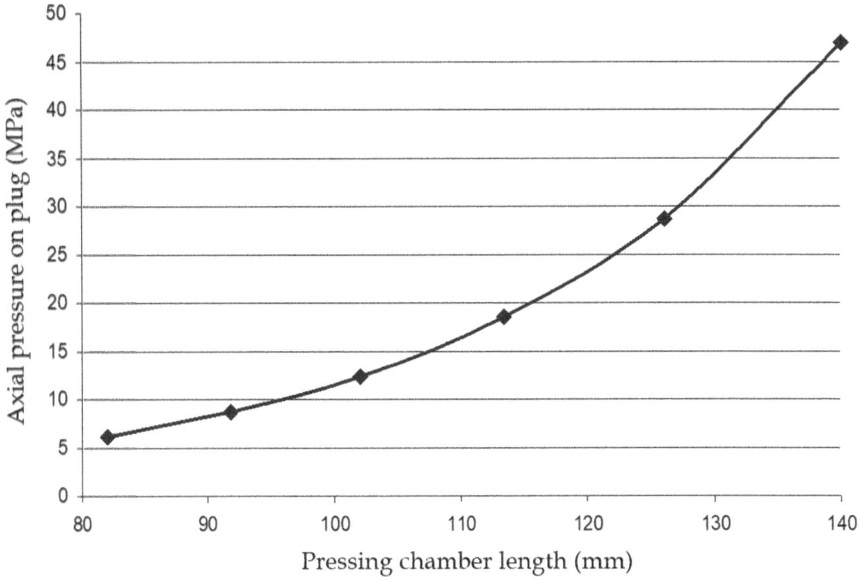

Fig. 5.3 Dependence of axial pressure on the extrusion "plug" with respect to chamber length

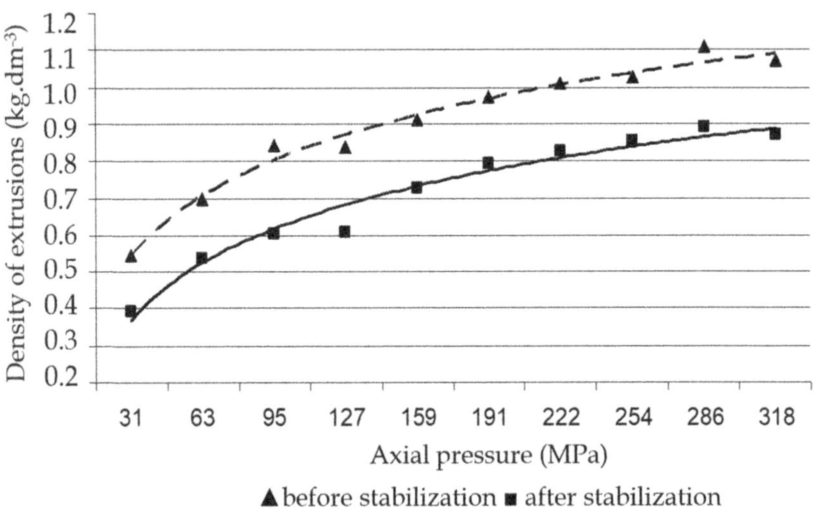

Fig. 5.4 Dependence of extrusion density on the axial pressure (pressing pressure) [14, 15]

information, it can be stated that the change in chamber length affects the resulting density of extrusions.

Another mathematical model was proposed in Ref. [16, 17], which deals with the action of pressures occurring in the pressing chamber in a pellet die and considered the compression of an elastic material (see Fig. 5.5). The derived relationship gives the dependence of plug pressure within the die on varying pressing pressures.

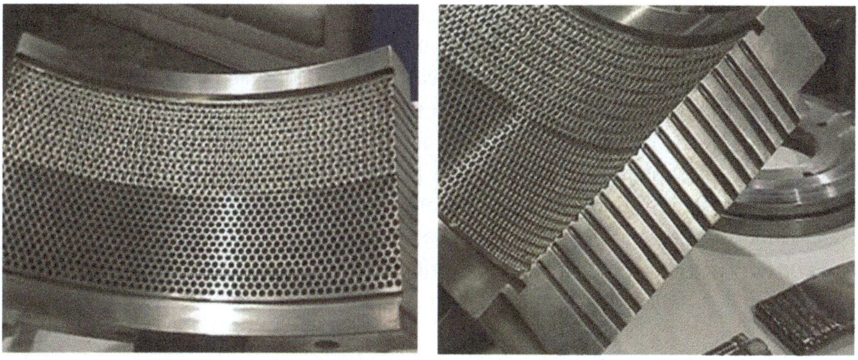

Fig. 5.5 Section of the horizontal roller pellet chamber and view of the die section

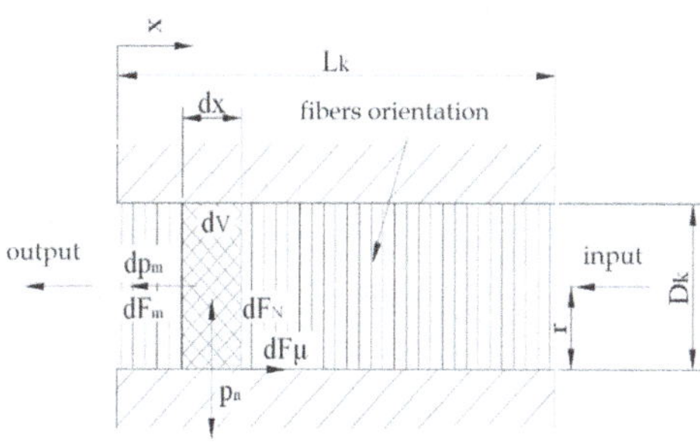

p_m -plug pressure (MPa)
F_m - plug force (N)
F_N - normal (radial) force (N)
F_μ - frictional force (N)
p_N - normal pressure (MPa)
D_k - diameter of the pressing channel (mm)
r - radius of the pressing channel (mm)
L_k - length of the pressing channel (mm)

Fig. 5.6 Schematic representation of forces and pressures acting on an element dV within the pressing channel [16, 17]

The selected element (see Fig. 5.6) has the shape of a flat disk with thickness dx and radius r. The center of this disk is identical to the axis of the nozzle. The differential of pressure change dp_m on the section dx is given by:

$$dp_m = \frac{dF_m}{\pi.r^2} \Rightarrow dF_m = dp_m.\pi.r^2. \tag{5.8}$$

The friction force dF_μ acting on the differential surface $2.\pi.r\,dx$ is given by

$$dF_\mu = \mu.dF_N = \mu.p_N.2.\pi.r.dx. \tag{5.9}$$

The resulting relationship was derived from the assumption that wood is an orthotropic material (the mechanical properties are independent of the directions parallel and perpendicular to the orientation of the fibers). The authors in Ref. [16, 17] assumed that the bundles of fibers in a pellet were oriented perpendicular to the largest dimension of the pressing chamber. Moreover, it is assumed that the material has elastic properties, i.e., the relationship between stress and strain applies (Hooke's law). In Ref. [16, 17], it is further assumed that the material is homogeneous throughout the entire cross-sectional area. In the case that the pellet is spatially restricting in the pressing channel, radial expansion cannot occur and thus a stress on the walls of the pressing channel manifests instead. The differential equation is then solved by substituting the relative pressures in the longitudinal and radial directions (p_r/p_k) introducing the conditions that p_r is equal to p_N and p_k is equal to p_m, considering Poisson's coefficient to be the ratio of unit strain for an element ν_{RL}, and considering Hooke's law for the longitudinal and radial directions of fiber bundles, we get

$$p_N = G.\upsilon_{RL} \cdot p_m \left(\text{MPa}\right), \tag{5.10}$$

where G is the ratio of the modulus of elasticity for the fiber bundles in the longitudinal and radial directions (E_L/E_R). If we then substitute Eqs. (5.8) and (5.9) into this equation, we get the differential equation for p_m

$$dF_m = dF_\mu \rightarrow dp_m.\pi.r^2 = \mu.p_N.2.\pi.r.dx. \tag{5.11}$$

Rearranging this equation by substituting Eq. (5.10) and solving for $x - x_0 = L_k$, the authors in Ref. [16, 17] obtained the following result, which describes the dependence of changing pressing pressure on the plug within the pressing channel of the die, where p_{m0} is the initial pressing pressure:

$$p_m = p_{mo} . e^{\frac{2.\mu.G.\upsilon_{RL} \cdot L_k}{r}} \left(\text{MPa}\right). \tag{5.12}$$

This mathematical model was used for numerical analysis to determine the influence of the pressing chamber length. In analyzing the effects of changing pressing chamber length, the same parameters were chosen as in the previous analysis. However, the parameters have been supplemented by the modulus of elasticity of fiber bundles for pine sawdust in the longitudinal and radial directions, as this model requires these data. Mechanical properties of pine are: $G = E_L/E_R \rightarrow E_L = 6634$ MPa; $E_R = 320$ MPa; $\upsilon_{LR} = 0.337$; $\upsilon_{RL} = 0.016$.

$$\frac{E_L}{E_R} = G.\upsilon_{RL} = \upsilon_{LR}. \tag{5.13}$$

In the previous Table 5.2, it is possible to observe calculated data. In the following Fig. 5.7, these data are expressed graphically. The results of this mathematical analysis show the influence of channel length within the pressing chamber, its

Table 5.2 Table of calculated results for different chamber lengths [12, 13]

L (mm)	**140**	126	113.4	102	91.8	82
Δ L (%)	–	−10	−10	−10	−10	−10
Pm (MPa)	**271.8**	195.4	145.1	111	87.2	69.2
Δ Pm (%)	–	−28	−25.7	−23.5	−21.4	−20.6

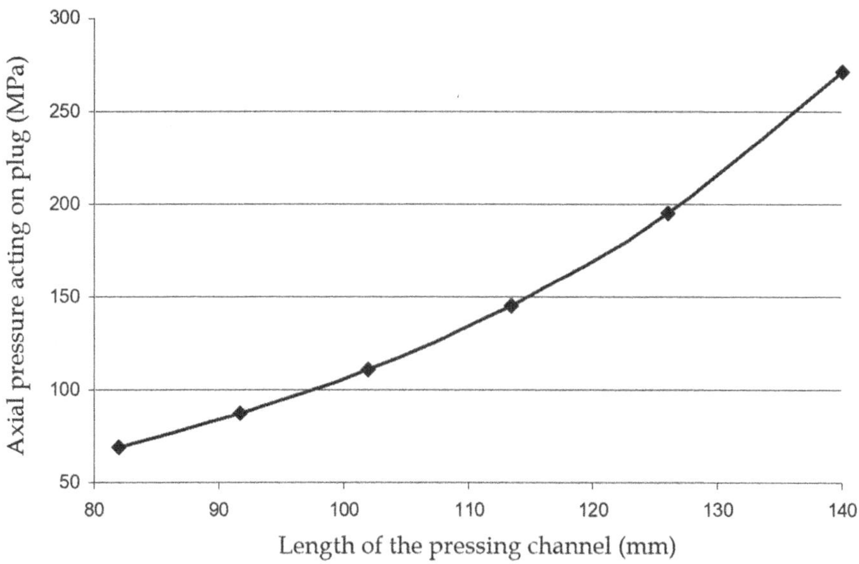

Fig. 5.7 Dependence of axial plug pressure on the change in channel length within a pellet die

behavior, and the action of pressures within the chamber and, consequently, also on the resulting quality of the extrusion.

The results obtained using the previous mathematical model can be supplemented by the results of experimental research, which was published by the authors in Ref. [18, 19]. The published experimental research was carried out on a pellet mill with a horizontal cylindrical die. In the experiment, the authors observed the effect of die thickness, which corresponded to the length of the pressing channels within the die. The results related to the mechanical resistance of the pellets, which is a typical indicator of the quality in pellets, can be seen in Fig. 5.8. The authors in Ref. [18, 19] used 3 cylindrical dies ranging in width from 29, 35, and 44 mm, respectively. Based on the results obtained, it is possible to support the presented assumption that changing the pressing channel length influences the final quality of the extrusions.

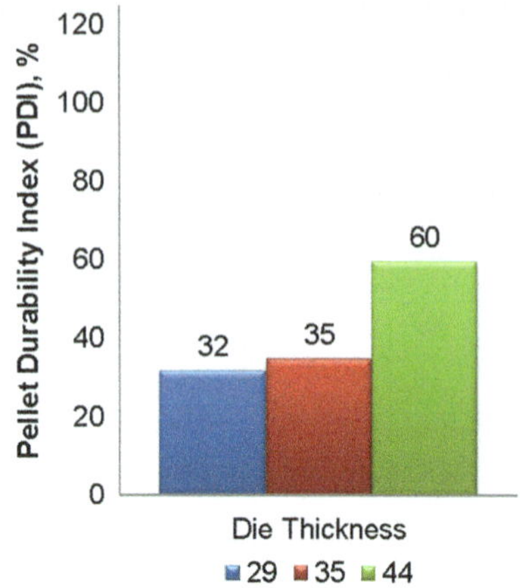

Fig. 5.8 Dependence of the mechanical durability of pellets on the thickness of the pellet die [18, 19]

5.2 Effects of the Pressing Chamber Diameter

A mathematical model (5.3) rewritten in the form (5.7) was used for a theoretical analysis investigating the effects of changing the diameter of the pressing chamber. The parameters remained unchanged throughout the analysis, with the exception that the length of the pressing chamber was set to 140 mm. When analyzing the effect of the change in diameter of the pressing chamber, the diameter $d = 20$ mm was chosen as an initial value, which was subsequently increased by 20% within the range <20 ÷ 50> mm (Table 5.3). These values for the diameter of the pressing chamber were chosen on the basis of possibilities provided by an existing pressing machine available at our institute.

The result of this analysis confirmed the assumptions that when changing the area (increasing the area) under a constant force, the magnitude of the pressure will also change (the pressure decreases). However, the purpose of this analysis was to determine the effects of changing the diameter of the channel. As the diameter of the pressing chamber increases, the axial pressure on the plug P_m decreases exponentially, which negatively affects the resulting quality of extrusions; see Fig. 5.9.

An interesting interaction worthy of experimental research is the investigation of the mutual interactions occurring between changes in the pressing channel diameter and its length; see Fig. 5.10. A mathematical program developed within the *Mathematica* programming environment was created for the abovementioned investigation. It is based on rewriting Eq. (5.7) into computer transcription, which allows for the solution of expected values with respect to selected parameters of the experiment.

Table 5.3 Table of calculated values for different diameters of the pressing channel [12, 13]

d (mm)	**20**	24	29.8	35.56	42.67
Δ d (%)	–	20	20	20	20
Pm (MPa)	**46.95**	21.1	9.5	6.15	3.48
Δ Pm (%)	–	55	55	35.3	43.4

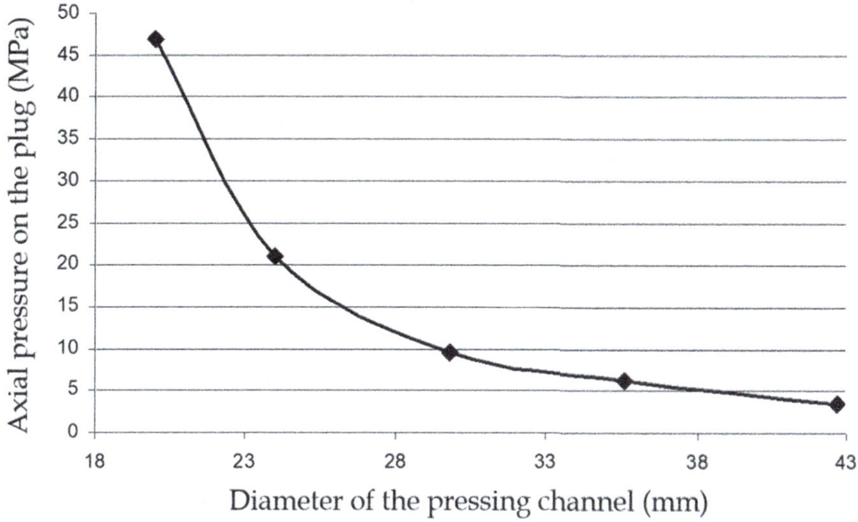

Fig. 5.9 Dependence of the axial plug pressure on the diameter of the pressing channel

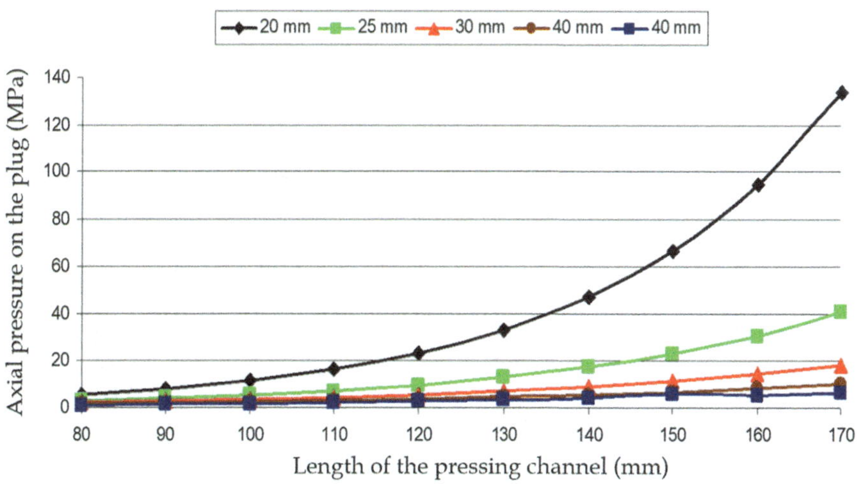

Fig. 5.10 Dependence of axial plug pressure on changes in channel length for different diameters of the pressing channel

5.3 Effects of Pressing Channel Conicity

As was mentioned above, the geometry of the pressing chamber is detrimental to the compaction of biomass. Based on practical experience with briquette production of wood biomass, we know that when changing the briquette nozzle, i.e., replacing the cylindrical nozzle for a nozzle with a particular conicity, positive effects on the resulting quality of extrusions were observed. This fact provides a clear assumption that there is a relationship between the degree of taper and pressing pressure on the resulting quality of extrusions. Therefore, it is important to determine the effects of changing the conicity of the pressing chamber nozzle and its effect on the resulting quality of extrusions. It is also possible to gain knowledge about the required dwell time of the extrusion within the pressing chamber, during which the extrusion is subjected to calibration where the axial and radial pressures are rearranging. As mentioned before, due to the higher residual radial pressures as the extrusions are being pushed out of the nozzle, extrusions may crumble and/or rupture. With the correct shape of the pressing chamber, we can prevent this.

Since there is no known mathematical model for such investigation, a modified version of the mathematical model (5.3) published within [9, 10] was used.

Figure 5.11 can be described by the following equation:

$$P_M.dS_V.\cos^2\alpha.S_{PL} - dP_M.S_{V2} - \mu.\left[\left(P_R + P_M.dS_V.\sin\alpha\right).S_{PL}\right]\cos\alpha = 0, \quad (5.14)$$

where

$$dS_V = 2.\pi.r.dr \quad (5.15)$$

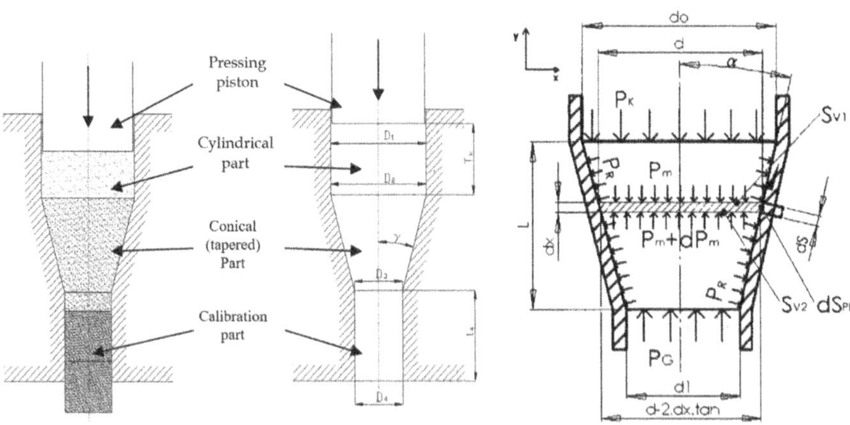

Fig. 5.11 Main parts of a tapered pressing chamber (left) and the distribution of pressures within the chamber (right), where p_k—axial piston pressure (MPa); p_G—chamber back pressure (MPa); p_r—radial pressure (MPa); p_m—axial plug pressure (MPa); μ—coefficient of friction (−); L—plug length (mm)

$$S_{PL} = \frac{\pi.(d_0 - d_1)}{2}.\frac{L}{\cos\alpha} \tag{5.16}$$

$$S_{V2} = \pi(r - d_r)^2 \tag{5.17}$$

By solving Eq. (5.14) for the boundary conditions: $x = 0$, $p_m = p_k$, for $x = L$, it is possible to describe the distribution of pressures during compaction in a conical pressing chamber, which is completely different compared to that of a cylindrical chamber. Even without the resulting mathematical expression, it is obvious at first glance that the resulting dependence describing the action of pressures in the chamber will have an exponential character. The main reason for this is that the taper of the chamber walls causes part of the axial pressing forces to be converted into a radial component, which causes higher frictional forces. On the other hand, pressing through a conical chamber causes compaction in the horizontal direction, not only along the pressing axis (the vertical direction in this case) [20, 21]. Therefore, it is possible to state, in order to optimize the geometry of the pressing chamber, both these variables. Increasing the taper angle of the chamber increases the pressure and thus the quality of extrusion; however, the friction between the material being pressed and the walls of the chamber and the associated adverse effects of its action are exacerbated (tool wear, chamber temperature, charring of the extrusion surface, etc.) [22, 23].

Since changing the taper in the pressing chamber influences the distribution of the applied forces and thus also the effects of wear in the pressing chamber, it is also possible to express the effect on the surface-to-unit volume ratio of the pressing chamber. For comparison, it is possible to mention 2 types of pressing chambers—conical pressing chamber (Fig. 5.12 on the right) and combined pressing chamber (Fig. 5.12 on the left), conical pressing part with cylindrical calibration part.

Input diameter is the same for both chambers $d = 20$ mm. The total lengths of both pressing chamber configurations are also the same at $l_h = 140$ mm. The length of the cylindrical part for the combined pressing chamber is constant $l_k = 60$ mm; i.e., the taper of the chamber is graduated at intervals of 1, 4, 6, 9, and 11° over the remaining length ($x = 80$ mm). In the case of a uniformly tapered pressing chamber,

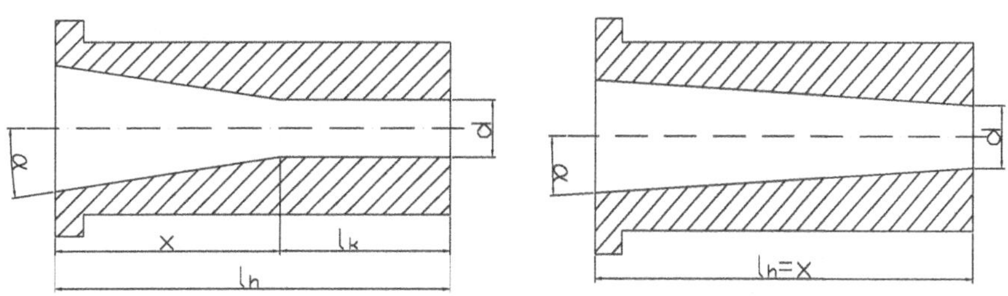

Fig. 5.12 Analyzed pressing chamber configurations (configuration "α H l_h/l_k" left, and configuration "α H l_h" right)

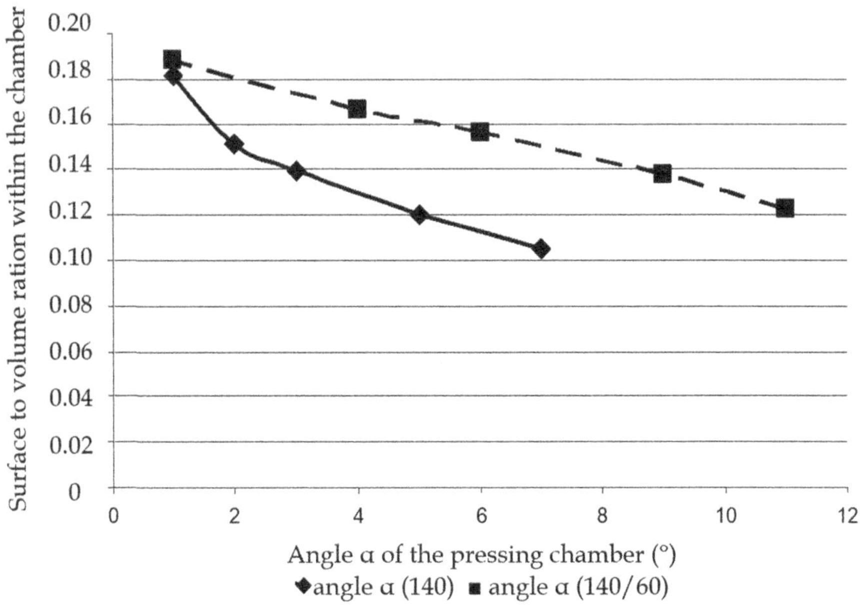

Fig. 5.13 Dependence of the surface-to-volume ratio on the pressing chamber taper angle

only the angle α will vary in intervals of 1, 2, 3, 5, and 7° over the entire length l_h = 140 mm of the chamber. As we can see in Fig. 5.13, surface-to-volume ratio is changing within the change in taper of the chamber.

5.4 Influence of the Coefficient of Friction

In the process of compacting, we encounter the requirement for compaction of various types of materials (softwoods, hardwoods, agricultural products, straw, paper, textiles, rubber pulp, cardboard, etc.). According to the type of mutual motion (tool—pressed material), we distinguish sliding friction throughout the process. Sliding friction is the result of a mutual displacement between two surfaces. Sliding friction manifests itself in the form of a friction force, friction torque, and mechanical frictional energy. The energy consumed in the friction process is converted primarily into heat, deformation processes, wear (creation of a new surface), structural changes, chemical reactions, and emissions of wear products. The friction and wear processes of the surfaces of a friction couple can be effectively eliminated by separating them with a layer of lubricant with suitable shear properties, which prevent their contact during their mutual sliding motion. However, in the process of compacting biologically "clean" materials for use primarily in the extraction of energy, common industrial lubricants are not permitted. Therefore, only certain types of biological additives (e.g., starch) can be added to the process for the reduction in friction. However, the permissible amount of biological additives is limited by technical standards for solid biofuels. Therefore, it is necessary to address the issue of mitigating friction conditions by changing the type of material used in the pressing tools with respect to the type of material being pressed or by designing the optimal shape of the pressing tools. This requires knowledge of the effect of the coefficient of friction on the resulting quality of the extrusions.

Based on the decomposition of forces acting within the pressing chamber, it can be seen that the friction force between the pressing tool and the pressed material is parameter that will significantly affect the durability and wear of the pressing tools. The influence of this parameter was subjected to mathematical analysis using the mathematical model (5.3) in Ref. [9, 10]. In analyzing the effects of changing the coefficient of friction, an initial value of $\mu = 0.1$ was chosen. This value was gradually increased by 20% within the range <0.1 ÷ 0.4>, which is the interval of the values for the coefficient of friction between steel and wood. The remaining parameters of the pressing process were chosen similarly to the analyses mentioned above and also take into account the dimensions of the experimental pressing apparatus used for measurements at our institute.

Based on the results (dependency shown in Fig. 5.14), it can be stated that the axial pressure required on the plug increases exponentially as the coefficient of friction increases. Also, by increasing the coefficient of friction, the axial pressure on the plug is also increasing nonlinearly, which directly affects the quality of extrusions.

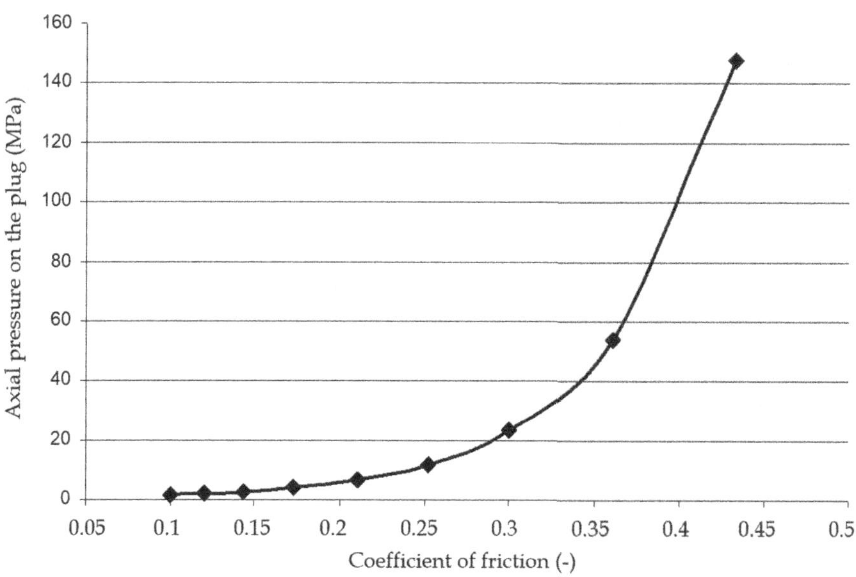

Fig. 5.14 Dependence of axial pressure acting on the plug on the change in the coefficient of friction

5.5 Effects of the Pressing Tool Shape

As mentioned above, pressing mill manufacturers have a wide portfolio when it comes to types of compaction tools. In the case of pellet mills, we always distinguish a co-related pair of pressing tools, i.e., the pressing tool and corresponding die (cylindrical, conical). The pressing rollers force the fed material through the die. Compaction occurs in the first step between the roll and the die, and subsequently within the pressing channels of the die (array of channels). The degree of compaction depends primarily on the type of material being pressed, but also on the kinematic couple of the machine and the shape of the pressing tools. However, pellet presses also use a method of deriving the pressing force by means of a screw pushing material through a plate die, as well as the principle of pressing between the intersect of two cylindrical dies. In each case, the most important geometric parameter is the shape of the pressing channels within the die. Different shape variations are used for the pressing channels within cylindrical or plate dies of pelleting mills; see Fig. 5.15. In the case of pellet mills, the subject of research deals with the specific shape of the pressing tools depending on the material being pressed. Press manufacturers have developed various pellet dies depending on the material being pressed. Thus, the back pressure sought is solved here exclusively by means of the shape pressing channel within the die.

In briquetting presses, we also distinguish between a co-related and working pair of pressing tools, pressing tool—nozzle. As mentioned before, the pressing tool can be a piston or a screw. The nozzle is in principle always cylindrical in shape, but manufacturers are gradually developing various modifications to the pressing nozzle, whereby they use a combination of cylindrical surfaces and conical surfaces. When producing briquettes on briquetting mills, primarily utilizing an open pressing chamber, it is necessary to ensure sufficient back pressure "p_G" over a defined duration (Fig. 5.16). This is also due to the pressing method itself, which is unstable from the point of view of ensuring a constant pressing pressure. In principle, the back pressure in the pressing chamber can be ensured by [9, 26]:

→ Controlled "throttling" the pressing chamber
→ Controlling the so-called TDC "Top dead center"
→ The shape of the pressing chamber

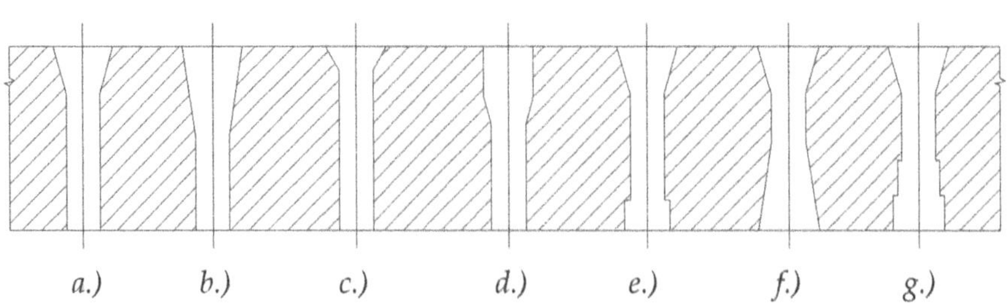

Fig. 5.15 Examples of different pressing channel shapes—(**a**) classical, (**b**) deep, (**c**) shallow, (**d**) well, (**e**) cylindrical, (**f**) conical, (**g**) stepped [24, 25]

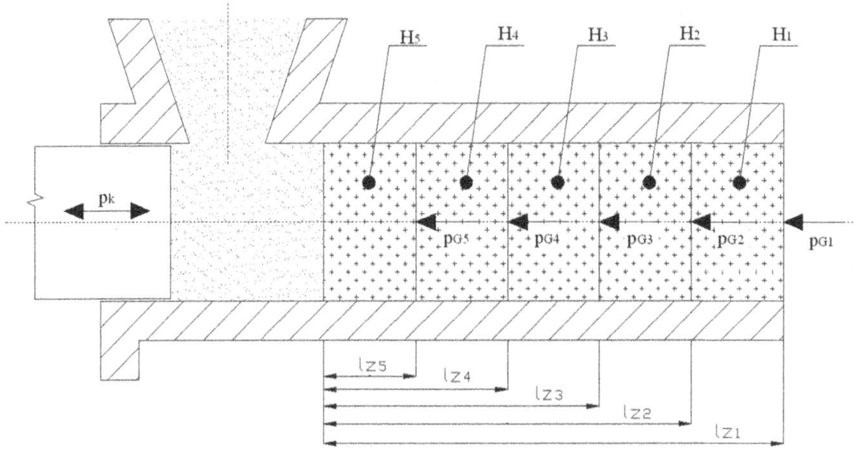

Fig. 5.16 Behavior of back pressure during continual compaction

The controlled "throttling" of the pressing chamber is primarily a method where the end of the material flow is exposed to a constant back pressure. This is derived by a throttling device, which acts as a valve. If the required pressing pressure deviates from the allowable tolerances (which affects the final quality of the extrusions), a "throttle valve" will regulate the process—thus causing a change in the back pressure. In most cases, the "throttling" is solved by means of an active weight on a hinged flap at the exit of the chamber or by constriction of the cooling channel, where the extrusions pass after compaction. "Throttling" can also be achieved mechanically, depending on the degree of automation, as well as hydraulically.

The "Top dead center" (TDC) control is based on the principle that any deviation from the desired back pressure value is eliminated by a short-term adjustment of the TDC point of the piston. It should be noted that this method is currently under development, precisely because there are several types of briquette press designs, and it is not possible to apply this method everywhere. With this method of control, the length of the pressed column in the chamber is changed and thus also the back pressure for the next pressing cycle. Therefore, plugs of material with a higher or lower degree of initial compression are shaped such that the pressure increases or decreases within the following cycle. The determining variable here is the constant back pressure $p_{G,}$ while variable(s) that adjust the "top dead center" of the piston are changed. Thus, there is an axial displacement of the piston housing. This control method assumes that the change in pressing pressure is unambiguously defined by a change in back pressure. It is therefore an indirect method of control.

Controlling the shape of the pressing channel is a method where, by means of adjustable channel geometry, changes in shape affect the radial lateral pressure p_r due to the action of the wall of the channel on the pressed material. This change is achieved by means of a "constricting" channel, which induced frictional resistances between the wall and pressed material so as to change the back pressure of the already pressed material for subsequent pressings. Most commonly, the mechanisms involved regulate the circumferential geometry of the free end of the pressing chamber, or by changing the shape of the nozzle(s). The first method can be implemented mechanically as well as in an automated manner, but when setting the conicity of the channel, there is often a limitation in the flexibility of the pressing chamber material itself. The second method cannot be implemented during the pressing process. When replacing the press nozzle(s), the machine must be shut down and cleaned, so these adjustments must be made before operation. Therefore, it is appropriate to have a pressing chamber composed of several segments (pressing inserts). These can have different shapes depending on the material being pressed and thus have a positive effect on the final quality of extrusions.

In any case, there is a higher rate of wear when compared to standard cylindrical extrusion chambers, so it is necessary to address the design of the shape of these extrusion chambers and the appropriate selection of material and heat treatment. A parameter that makes it possible to define in advance at least an approximate degree of wear of the pressing chambers is the surface-to-unit volume ratio of the pressing chamber. The lower this ratio, the lower the assumption that the press chamber will undergo accelerated wear. Since wear is also related to the distribution and action of pressures within the pressing chamber, it can be assumed that the shape of the pressing chamber also affects the resulting quality of extrusions. The goal of the following theoretical analysis was to represent the influence of the shape of the pressing chamber in the pressing process through the surface-to-unit volume ratio of the pressing chamber.

When comparing the nozzles in terms of the calculated surface area and the calculated volume, it is clear that when using different types of nozzles (Table 5.4, Fig. 5.17), we achieve different hourly production of the machine with the same input energy. Figure 5.18 shows a comparison of nozzles in terms of surface-to-volume ratio. The 7H140 nozzle appears to be the most suitable in terms of surface-to-volume ratio. This configuration displays the least wear, which is economically advantageous. The analysis shows that it is suitable to use tapered nozzles in the pressing chamber along the length of the pressing chamber. As these nozzles wear the least, frequent nozzle changes are not necessary, and thus, the productivity may be higher than with other types of nozzles. It can be noted that nozzles with a cylindrical pressing chamber along the entire length of the pressing chamber appear to be the most unsuitable types of pressing nozzles from the point of view of extrusion quality. Nozzles with certain conicity are more suitable, due to the action and distribution of pressures in the pressing chamber.

The data in the figure above are obtained by calculation and consider rather the dimensions of the pressing chambers, which could be applied on the experimental pressing mill at our institute. Of interest, however, it can be observed when comparing commercially available pressing chambers. This comparison can be observed in the following Fig. 5.19. Here, you can see a comparison of the surface-to-volume

Table 5.4 Designed nozzle(s) dies used throughout the theoretical analysis [27, 28]

Nozzle designation	Length of the nozzle l_h (mm)	Nozzle length with output diameter l_k (mm)	Input diameter *d* (mm)	Taper angle α (°)	Length of the tapered part *x* (mm) ($x = l_h - l_k$)	Nozzle volume S_k (mm^2)	Nozzle volume V_k (mm^3)	Surface-to-volume ratio S_k/V_k (mm^{-1})
H 140	140	–	20	–	–	8922	43982	0.203
H 120	120	–	20	–	–	8168	37699	0.217
H 100	100	–	20	–	–	6912	31416	0.220
H 70	70	–	20	–	–	5027	21991	0.229
H 140/120	140	120	20	39	20	13501	59397	0.227
H 140/100	140	100	20	17	40	12945	65094	0.199
H 140/70	140	70	20	7	70	12557	66267	0.189
7 H 140	140	–	20	7	140	18374	151886	0.121
5 H 140	140	–	20	5	140	15960	117873	0.135
3 H 140	140	–	20	3	140	13667	88551	0.154
2 H 140	140	–	20	2	140	12564	75650	0.166
1 H 140	140	–	20	1	140	10309	53365	0.193
11 H 140/60	140	60	20	11	80	16062	105642	0.152
9 H 140/60	140	60	20	9	80	14365	86206	0.167
6 H 140/60	140	60	20	6	80	12802	69451	0.184
4 H 140/60	140	60	20	4	80	12069	62078	0.194
1 H 140/60	140	60	20	1	80	10557	49344	0.214

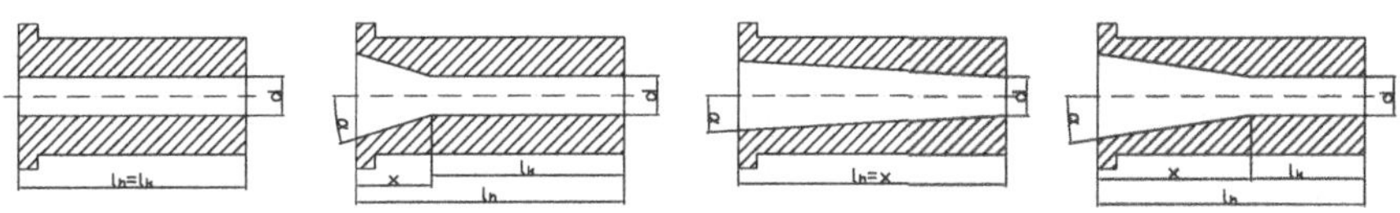

Fig. 5.17 Schematic representation of different nozzle configurations (from left nozzle "H l_h" ; nozzle "H l_h/l_k" ; nozzle "α H l_h"; nozzle "α H l_h/l_k")

ratios in the pressing channels of pellet dies (plate dies) with diameters of 6, 8, and 10 mm, and briquetting nozzles with diameters of 50 and 90 mm. This comparison is complemented by a briquette nozzle, which was developed at our institute (see Fig. 3.33), with which it is possible to produce 4 extrusions, 20mm in diameter simultaneously, hence the designation 20/4.

It is possible to recognize significant differences between the pellet and briquette pressing chambers, and these results also confirm the higher wear of pellet dies compared to briquette chambers. On the other hand, from the point of view of extrusion quality, in principle, it can be said that pelleting machines can produce extrusions of higher quality than briquette machines.

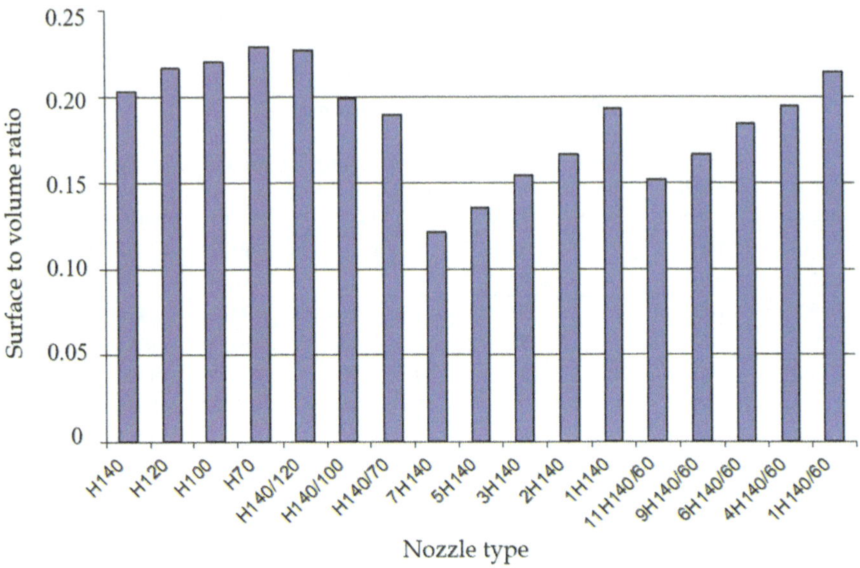

Fig. 5.18 Comparison of different nozzle types in terms of their surface-to-volume ratio [27, 28]

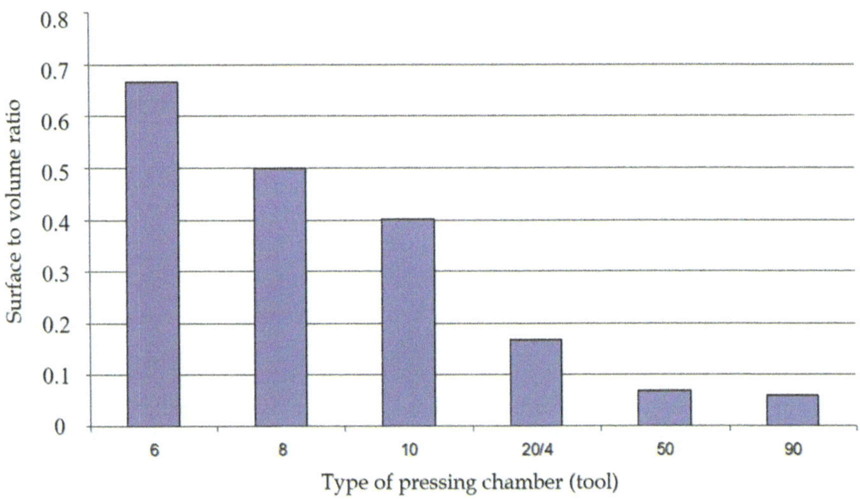

Fig. 5.19 Comparison of different commercially available pressing chambers (tools) [27, 28]

5.6 Summary of the Theoretical Analysis

The analyses performed show that the design parameters significantly affect the pressures within the pressing chamber and, as a result, also affect the resulting quality of extrusions. For some of the parameters mentioned (on page 75), it was not possible to perform a separate theoretical analysis, either due to the lack of adequate information or due to the interconnectedness and interaction of these parameters with other analyzed parameters. Therefore, the influence of the length of the cooling channel, the influence of the pressing method, or the influence of the material of the pressing tools was not included in the theoretical analysis. However, based on practical experience, it is important to deal with the experimental verification of the influence of all of the above parameters. On the other hand, based on the available experimental pressing equipment, it is not technically possible to experimentally verify all of the analyzed parameters. Therefore, it was necessary to select parameters appropriately with regard to the size and significance of their influence, particularly on the quality of extrusions, with regard to the time and economic complexity of the experiment and with regard to limitations of available experimental equipment. After thorough analysis and due to the possibilities of the experimental pressing stand, only the following design parameters were selected for experimental verification:

- Length of the pressing chamber
- Conicity of the pressing chamber walls
- Pressing method
- Distribution and action of frictional forces within the pressing chamber

The benefit of the analysis and study of design parameters and their influence on the process is the newly acquired knowledge, which can be applied to future experimental apparatus allowing for possible experimentation to verify the influence of all design parameters. Therefore, one of the benefits of the findings within this monograph is information, which will be used for the design of a new experimental pressing apparatus, which will allow for the monitoring of:

- Diameter of the pressing chamber
- Type and geometry of the pressing chamber
- Material of the pressing tools
- Applied frictional forces, radial forces, and back pressures
- Length of calibration (cooling) channel

References

1. En Plus Handbook; European Pellet Council: Handbook for the Certification of Wood Pellets for Heating Purposes v.2.0, based on EN 1496, 1–2, April 2013
2. Holm, J.K., et al.: Fundamentals of Biomass Pellet Production. Annual Research Report (2005). ISBN 87-74-75325-8
3. Holm, U.B., et al.: Fundamental Understanding of Pelletization. EFP-2005 Project (33031-037) – Summary Report, p. 51. Lyngby, Denmark (June 2008)
4. Križan, P., Matúš, M., Svátek, M.: Findings from the research of pelleting of selected species of dendromas and phytomasses. In: Briquetting and Pelleting 2012: A Collection of Lectures from the 6th International Conference. Bratislava, 25.-26.1. 2012, pp. 104–111. STU in Bratislava, Bratislava (2012). ISBN 978-80-227-3641-1 (in Slovak)

5. Križan, P., Šooš, Ľ., Vukelič, D.: Counter pressure effecting on compacted briquette in pressing chamber. In: MMA 2009. Flexible Technologies : Proceedings. 10th International Scientific Conference. Novi Sad, 9–10.10. 2009, pp. 136–139. Faculty of Technical Sciences, Novi Sad (2009). ISBN 978-86-7892-223-7
6. Križan, P.: Research of factors influencing the quality of wood briquets. Acta Montanistica Slovaca. **12**(3), 223–230 (2007). ISSN 1335-1788
7. Chudý, V., et al.: Measurement of Technical Quantities, p. 438. STU Bratislava (1999). ISBN 80-227-1275-2 (in Slovak)
8. Meloun, M., Militký, J.: Statistical Analysis of Experimental Data. Academia – Publishing House, Praha (2004). ISBN 80-200-1254-0, 953 pp (in Czech)
9. Horrighs, W.: Control of ram extruders for obtaining steady compaction pressures. J. Prep. Process. Aufbereitungs-Technik, nr. 07/1987, pp. 378–382, Wiesbaden (1987). ISSN 0004-783 X
10. Horrighs, W.: Determining the dimensions of extrusion presses with parallel-wall die channel for the compaction and conveying of bulk solids. J. Prep Process. Aufbereitungs-Technik, nr. 12/1985, pp. 724–732, Wiesbaden (1985). ISSN 0004-783 X
11. Križan, P.: The Process of Pressing Wood Waste, 1. vyd, 197 p. Publishing House STU, Bratislava (2014). ISBN 978-80-227-4251-1 (in Slovak)
12. Kováč, P.: Research into the Influence of Design Parameters on the Final Quality of Moldings. Master thesis. ÚSETM. FME STU v Bratislave, SjF-5254-26779, Supervisor: Peter Križan, 83 p (2010) (in Slovak)
13. Liška, M.: Research into the influence of design parameters on the final quality of compacts during compaction. Master thesis. ÚSETM. FME STU in Bratislava, SjF-20736-58456, Supervisor: Peter Križan, 91 p (2013) (in Slovak)
14. Križan, P., Matúš, M.: Compacting pressure and pressing temperature at compacting process. In: Energetika a životní prostředí 2008 : Sborník přednášek z mezinárodní vědecké konferenci. Ostrava, 11.-12.9. 2008, pp. 77–82. Vysoká škola báňská – Technická univerzita v Ostravě, Ostrava (2008). ISBN 978-80-248-1832-0
15. Križan, P., Šooš, Ľ., Matúš, M.: Influence of pressing pressure and temperature during briquetting on dilatation of biomass compacts. In: Energetika a životní prostředí 2011: Proceedings of an International Scientific Conference. Ostravice, 29–31.8. 2011, pp. 84–92. Vysoká škola báňská – Technická univerzita v Ostravě, Ostrava (2011). ISBN 978-80-248-2456-7 (in Slovak)
16. Holm, J.K., et al.: Toward an Understanding of Controlling Parameters in Softwood and Hardwood Pellets Production, published on web 09/09/2006, American Chemical Society
17. Holm, J.K.: Pelletising different materials – an overview. In: Proceedings of International conference World Sustainable Energy Days, 1.-3.3. 2006, Wels, Austria
18. Stark, C.R., Ferket, P.: Conditioning, Pelleting and Cooling. citované online 24.08 2020 12:30, College of Agriculture and Life Science; http://www.ncsu.edu/project/feedmill/presentations/WS%202011/PDF2011/Conditioning%20Pelleting%20Cooling%202011.pdf
19. Stark, C.R.: Effect of die thickness and pellet mill throughput on pellet quality. In: Southern Poultry Science Society Meeting (2009)
20. Križan, P., Vukelič, D.: Shape of pressing chamber for wood biomas compacting. In: Quality Festival 2008: 2nd International Quality Conference. Kragujevac, May 13–15, 2008. University in Kragujevac, Kragujevac (2008). ISBN 978-86-86663-25-2
21. Tlacháč, P.: Design of Laboratory Compaction Equipment with Horizontal Pressing Axis, Master thesis. ÚSETM, FME STU in Bratislava, SjF-20736-50278, Supervisor: Peter Križan, 144 p (2012) (in Slovak)
22. Križan, P., et al.: Change of pressing chamber conicalness at briquetting process in briquetting machine pressing chamber. Acta Polytechnica. **52**(3), 60–65 (2012). ISSN 1210-2709
23. Nielsen, N.P.K. Gardner, D.J. Poulsen, T.: 2009. Importance of temperature, moisture content, and species for the conversion process of wood residues into fuel pellets. Wood Fiber Sci., 41(4), p. 414–415

24. Lehtikangas, P.: Quality properties of pelletised sawdust, logging residues and bark. Biomass Bioenerg. **19**, 351–360 (2001)
25. Perez, E.: Module – Pellets Press Components (Rolls, Dies, Knives, Feeder). The Agricultural University of Norway, online, cited 05.12.2020. http://docplayer.net/21050493-Module-pellet-press-components-rolls-dies-knives-feeder.html
26. Rumpf, H.: Die Wissenschaft des Agglomerierens, Chemie-Ingenieur-Technik, Nr. 1, pp. 1–11 (1974)
27. Križan, P., Šooš, Ľ., Vukelić, D.: Compacting of pine sawdust. J. ERIN (Edu. Res. Innov.). **1**(1), 2–7 (2008). ISSN 1337-9089
28. Križan, P., Vukelič, D.: Shape of pressing chamber for wood biomass compacting. Int. J. Qual. Res. **2**(3), 205–209 (2008). ISSN 1800-6450

Chapter 6
Experimental Research

Thorough data analysis is an inevitable part of experimental research. The need to determine the influence of individual design factors on the compaction process leads us to the steps necessary for research and the design of the experiment. It is therefore necessary to thoroughly prepare and plan for the experiment, its design, preparation of materials and instrumentation, implementation of research, followed by the subsequent processing and evaluation of data. It should be borne in mind that each experiment should have some informative value and a clear result (whether positive or negative).

Each measurement or experimental research consists of several consecutive steps [1, 2]. In order for the results of the experiment to be considered correct, the following steps must be followed throughout the experimental research:

1. Determine the goal of the experiment (measurement)—in the beginning, it is necessary to determine what the goal of the experiment will be and what parameter we will monitor in the result and how we will measure it (directly or indirectly—through another parameter).
2. Design the experiment (measurements)—design the experiment with respect to the number of monitored parameters, the amount of available data, and time and cost of measurements and prepare a thorough plan for its execution.
3. Prepare the experiment (measurement)—prepare materials used for experiments and preparation of instruments and equipment—their availability and functionality.
4. Execution of the experiment (measurement)—obtain the necessary information—data. The experiment plan must be followed. Measurement accuracy must also be observed.
5. Process the measured data—process the measured data and prepare them for further processing (different between calculations). Subsequently, verify the usability of the measured and calculated data for further processing and model creation. The procedures for this testing are given below.

P. Križan, *Biomass Compaction*, https://doi.org/10.1007/978-3-030-89956-1_6

6. Interpret the results with respect to the goals—correct interpretation of the results.

6.1 Design of the Experimental Research

6.1.1 Define the Experiment Goals

The basic task in designing an experiment and experimental plan is to specify, as clearly as possible, the goals. This means that it is necessary to determine exactly the meaning of the experiment, the observed variables in the experiment, the nature and type of output of the experiment, the method of data acquisition, and so forth. The entire "flow of information" and data (e.g., from material preparation to the creation of functional dependencies or mathematical model) must be predefined. Often times during the experiment, there is no room for significant changes in plan. Certain deviations made from the established experimental plan may trigger an unexpected "chain reaction" of undesirable circumstances. On the other hand, it can be said that there may be cases where it is necessary to make changes to the experimental plan during the experiment. In this case, it is important to take the necessary measures in favor of the usability of the data already obtained, the correctness of the experiment and with regard to the goals of the experiment.

In the field of biomass compaction process research, the primary goal is to gain knowledge about the behavior of important parameters, their impact on various aspects and factors of the biomass compaction process, and the effect of these on the compaction process itself [2, 3]. Based on experience and after detailed analysis of the impact of parameters on the biomass compaction process, we decided to determine the impact of the following design parameters on the compaction process:

1. Influence of the pressing chamber length
2. The influence of pressing chamber conicity
3. The influence of the pressing method
4. Distribution and action of friction forces in the pressing chamber

By determining the basic dependencies and the influence of the mentioned parameters, it is possible to obtain new knowledge, which can be used to optimize the pressing chamber for biomass compaction. Of course, it is never possible to design and perform an experiment to record the influence and behavior of all (or a larger group) of these parameters at once. Therefore, it is important to design the goal of the experiment very sensitively, i.e., to divide the monitored parameters into smaller groups and thus design several smaller experiments. However, if it is necessary to compare the results of smaller experiments with each other, or to interpret them together, it is important that there is a reasonable connection between the individual groups and experiments. In this case, it is necessary to relate all separate experiments to the final quality of extrusions, which is mostly defined by the density of the extrusions.

6.1.2 *Selection and Design of the Experimental Conditions*

The aim of this experimental research is not to propose a mathematical model that would describe the influence of the monitored parameters on the resulting density of extrusions, but to determine the influence of each of the monitored parameters individually. The selection and design of the experiment must be based on the conditions available for the experiment. These are the availability of individual data, the possibility of repeating measurements, environmental impacts, and so forth.

All experiments were performed on an experimental pressing apparatus available at our institute (see Fig. 6.1). This apparatus was designed and manufactured for the needs of experimental research to investigate the influence of technological and material parameters on the pressing process. This means that it is of a very simple construction, where a heating coil is installed allowing for smooth control in the pressing chamber temperature (see Fig. 6.2). The construction of the apparatus allows for only vertical pressing, which is performed on a hydraulic press.

The compaction of the samples took place on a hydraulic press, which works within a range of 10, 40, and 100 kN. The press makes it possible to regulate the force, which is very important in determining the pressing pressure during compaction. By simply converting the area on which the press piston acts, a pressing pressure from 31–318 MPa may be derived. The hydraulic press has a recording device (see Fig. 6.3), the advantage of which is the possibility to record the behavior of the pressing forces with respect to the position of the pressing piston within the chamber.

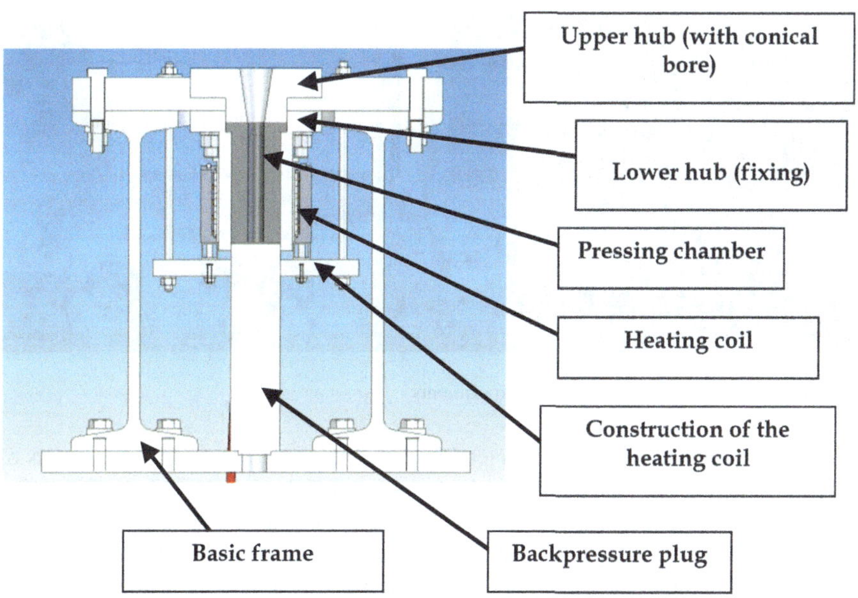

Fig. 6.1 Model of the experimental pressing stand and its basic components [4, 5]

Fig. 6.2 Experimental pressing stand prepared for experiments

Fig. 6.3 Hydraulic press used for the experiments

Based on the monitored parameters, it was necessary to prepare various parts of the experimental apparatus, mostly the pressing chamber—pressing nozzle(s). For the purpose of determining the effect of the chamber length, it was important to design the segment lengths of this parameter in which the experiment will take place. When designing the segment lengths, it was necessary to consider the limitations of the working space of the hydraulic press and also the experimental apparatus. The following Fig. 6.4 shows a section through the pressing chamber of the experimental stand. The maximum dimension, i.e., the length of the pressing chamber is 140 mm. The smallest usable length of the pressing chamber is 80 mm, due to the installation of a temperature sensor for the regulation of the heating coil. Therefore, the following chamber segment lengths were proposed for the experiment.

Since the pressing method of the hydraulic press is in the vertical orientation, it was necessary to design and manufacture back pressure plugs for all lengths of pressing chambers. The function of the back pressure plugs is, on one hand, to prevent the pressed material from falling out of the chamber after filling and, on the other hand, to cause back pressure during the pressing process. After initial compaction, the input material becomes pressed within the chamber and the back pressure plug can be removed allowing the extrusion to be ejected from the pressing chamber. The following Figs. 6.5 and 6.6 show the manufactured parts of the experimental stand for the needs of the experiment.

Since we must define the effect of the pressing chamber conicity, it was also necessary to define levels of this parameter for the experimental research. Based on the analysis performed for the influence of conicity and the limitations resulting from the construction of the experimental pressing stand, 3 levels for conicity were proposed (see Table 6.1). These angle values were designated based on the assumption that applied pressure increase within the chamber is exponential in comparison with the cylindrical chamber taking into account the maximum possible derived pressing forces of the press. The conical pressing chambers are fitted to the lower part of the pressing stand (see Fig. 6.7). This arrangement of the pressing chambers has been designed with regard to the dimensions of the pressing stand and the pressing piston (see Fig. 6.8). During the compaction process, the piston extends to the edge of the pressing chamber and the pressed material will be forced into the conical part of the pressing chamber. This method is also advantageous because it perfectly simulates the pressing process on production pressing mills.

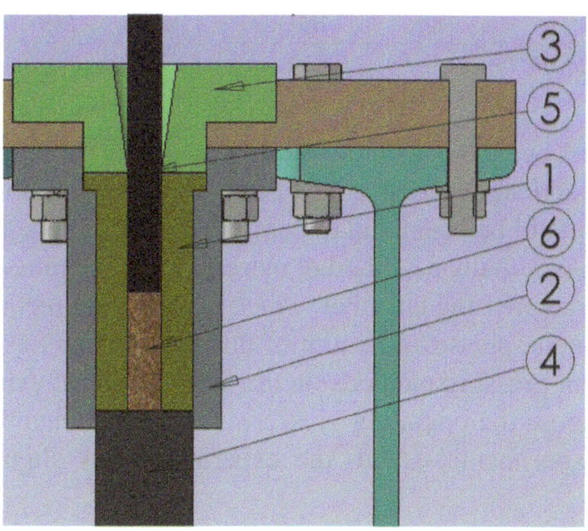

1 – pressing chamber
2 – pressing chamber hub
3 – input hub with conical bore
4 – backpressure plug
5 – pressing piston
6 – compacted material

Fig. 6.4 Section of the pressing chamber of the experimental pressing stand [6, 7]

Fig. 6.5 Interchangeable pressing chambers of different lengths

Fig. 6.6 Back pressure plugs of varying lengths

Table 6.1 Designated chamber length segments

Length of the pressing chamber (mm)	80	100	120	140
Pressing chamber conicity (degree of taper)	1	2	3	

The experimental pressing stand, available at our institute, was used in the experiments described. But based on practical experience, and considering the possibilities of the experimental stand, it would be beneficial to compare and determine the impact of an open chamber, which is actually a method of mechanical crank presses, and the effect of pressing in a closed pressing chamber, which is a pressing method that takes place in hydraulic briquette presses. The aim of this experiment was to determine the influence of different pressing methods on the final quality of extrusions and to obtain information about the operating pressures within the chamber. Due to the conditions of the experimental stand, the experiment was slightly simplified.

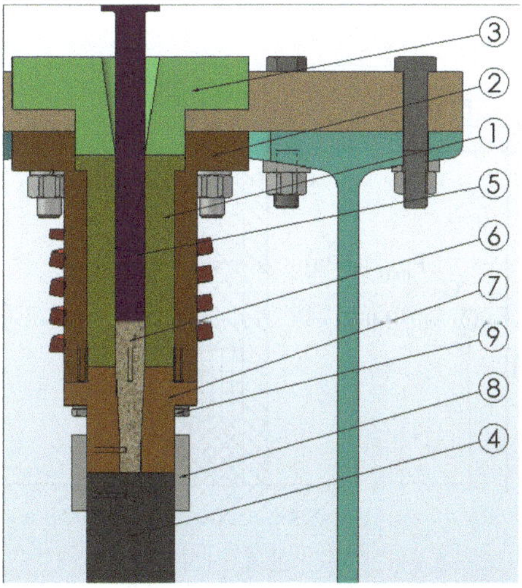

1 - pressing chamber
2 - hub of the pressing chamber
3 - Input nozzle with conical bore
4 - backpressure plug
5 - pressing piston
6 - pressed material
7 - conical pressing chamber
8 - collar
9 - bolt

Fig. 6.7 Section of the pressing stand with conical pressing chamber [6, 8].

Fig. 6.8 Conical pressing chambers

Compaction in the "open" pressing chamber takes place with a continuous supply of material into the chamber. The material being pressed is cyclically subject to an axial force, and thus, it is gradually compacted into solid extrusions as they are forced through the entire pressing chamber until the briquette is ejected from the chamber. The principle of this "open" pressing chamber can be observed when pressing several extrusions (Fig. 6.9); i.e., the chamber is filled several times (2, 3, and 4x) before pressing. The back pressure was released during pressing only when the extruded "plugs" had achieved sufficient back pressure. In essence, this system is generally used in "continuous" pressing chambers of briquette mills, i.e., chambers where no back pressure is created by closing the chamber completely. With such a simplified system, it is possible to monitor the compaction of individual plug

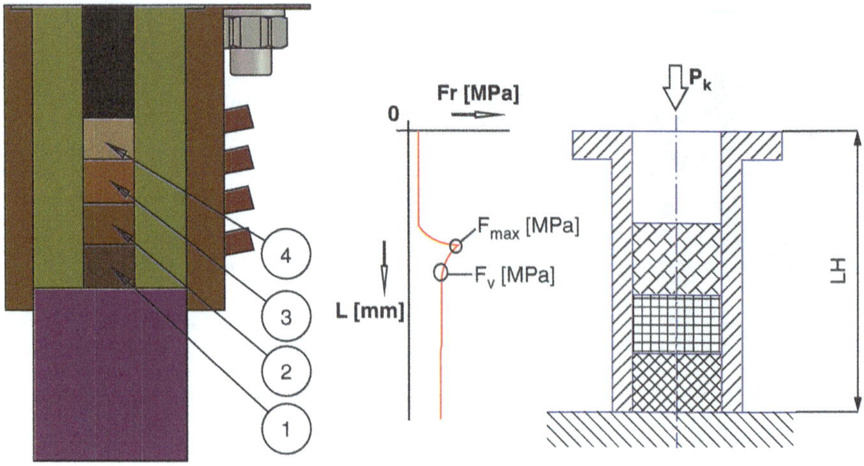

Fig. 6.9 Scheme of the pressed segment over multiple fillings—principle of an "open" pressing chamber [6, 8]

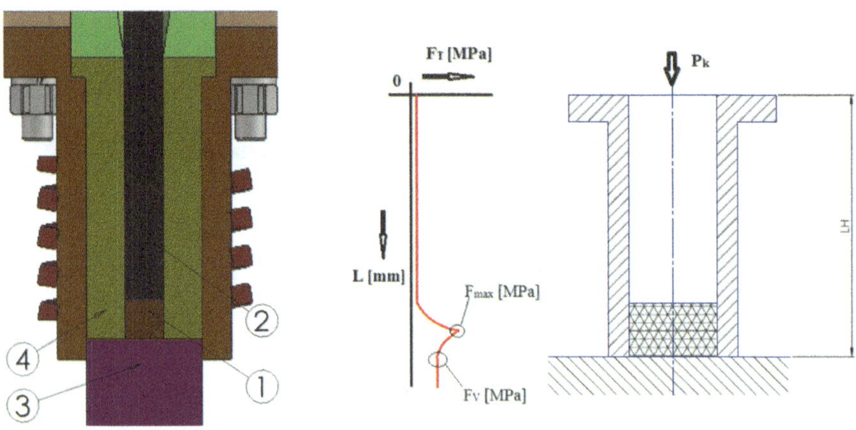

Fig. 6.10 Scheme of the pressing over one filling—principle "closed" pressing chamber [6, 8]

pressings within the chamber, their mutual influence, the action of the radial pressures, and frictional forces between the pressed material and the walls of the chamber.

Compaction in a "closed" chamber takes place within a closed space after filling with material. The pressing piston performs compacting in a closed cycle. This cycle includes a rest phase (time endurance) of the pressing piston, during which either a new material is fed into the chamber or the compacted briquette is ejected from the pressing chamber. The principle of a "closed" pressing chamber (Fig. 6.10) can be observed when producing a single extrusion per cycle; i.e., the chamber is filled with a dosage of the same (volume) of material before compaction commences. The bottom plug is released only after the extrusion is pressed through the chamber.

6.1.3 Pressing Procedure and Apparatus

The goal of the experiments was to monitor the quality of extrusions depending on the factors investigated. The quality of extrusions was evaluated in terms of its density. Due to the complexity of the experiments, only one type of material was used—pine sawdust. The experiments were performed on an experimental pressing stand, which was designed for the needs of the experiment. Due to the correct interpretation of the experimental results, it was necessary to produce 7 extrusions at each of the proposed levels [9]. This number was chosen for the purpose of future evaluation of the experimental results and for further processing of the measured data using known mathematical–statistical methods (analysis of means, analysis of variance, method of effects, etc.). In Ref. [10, 11], the literature indicates a minimum number of 6 values at one setting level. The experimental procedure for compacting pine sawdust is as follows:

1. Preparing the Input Material

Based on experience and previous experimental research of the influence of material parameters, we know that material properties (moisture and fraction size) have a significant impact on the final quality of the extrusions and affect the pressures acting within the chamber. In order to obtain correct results and determine the influence of the monitored parameters, it was necessary to properly prepare the input material. In order to isolate environmental factors, it was necessary to provide various fraction size and moisture content of the input material. To ensure different fraction sizes, a sieve analysis was performed [12], using an AS 200 screening device and circular sieves with mesh sizes of 0.5, 1, 2, and 4 mm (Fig. 6.11). In order to ensure different levels of moisture content within the input material, a laboratory drier BINDER ED 23 (Fig. 6.12) was used. Specific values were monitored and controlled by KERN MRS 120-3 humidity balancer (Fig. 6.13) and GMH 3830 digital hygrometer (Fig. 6.14) [13].

2. Dosage and Filling of the Pressing Chamber

Since the volume of the pressing chamber is known and therefore also the compaction ratio, it is possible to determine the dosing amount of sawdust required to fill the chamber. It was especially important to consistently add the same weighed amount of material to chambers, which produced a single extrusion. The pressing chamber was cylindrical with a diameter of 20 mm in a vertical configuration. When filling the pressing chamber with sawdust, it was necessary to use a back pressure plug, which prevented the sawdust from "leaking" out of the chamber during compaction. On the following Fig. 6.15 can be seen the filling area of pressing stand.

Fig. 6.11 Laboratory sieving apparatus, AS 200

Fig. 6.12 Laboratory drier

Fig. 6.13 balancer for measuring moisture content

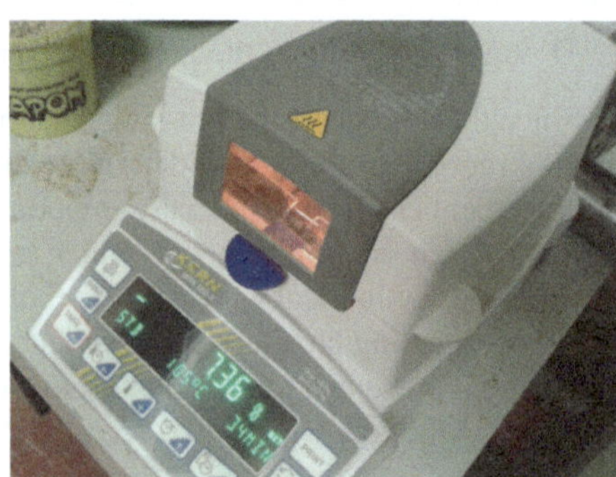

Fig. 6.14 Digital hygrometer

Fig. 6.15 Filling area of the experimental pressing stand

3. Pressing Procedure

After preparing the material in terms of fraction size and moisture content, compaction could take place. Throughout the compaction, it was necessary to follow the proposed levels of the monitored parameters and experimental conditions. The pressing device was also equipped with a heating coil, which can be used to regulate the temperature inside the pressing chamber. However, in determining the influence of the monitored parameters, it is not necessary to add heat, as this is counterproductive. Higher pressing temperatures cause the lignin within the input material to plasticize, which results in the conditions of friction within the pressing chamber to change. If we are to unambiguously investigate the influence of the monitored parameters, it is necessary that the compaction takes place without additional heating of the pressing chamber. In the next phase of the experiment, however, the heating of the pressing chamber is used. After filling the chamber with sawdust, compaction commences with pressing pressure(s) according to the determined experimental conditions. Depending on the type of experiment, after releasing the pressing pressure, the chamber can be refilled and the process repeated, until the pressed material is extruded from the pressing chamber (see Fig. 6.16).

4. Determining the Dimensions and Weight of Extrusion, Stabilization of Extrusions

After the pressed material is extruded from the chamber, the diameter of the extrusion was measured with a digital caliper 1 and the length of the extrusion with a digital caliper 2 (Fig. 6.17a, b). The use of two calipers, each for a separate dimension, is to avoid correlations of measured values that could occur in the calculation of measurement uncertainties [14, 15]. The weight of the extrusion was determined using a digital scale (Fig. 6.17c). Subsequently, the extrusion was placed under stable conditions and left to undergo the effects of dilation, i.e., extrusion stabilization. After the stabilization period, the dimensions (diameter and length of the extrusion) and the weight of the extrusion were again measured, and these values were the basis for determining the density according to the technical standard [16]. The densities obtained with one setting of values according to the experimental plan were averaged. This procedure was repeated for each of the determined levels of experimentation. Each extrusion was clearly identified with a unique marking, which during the entire course and processing of the experiment belonged to only one, specific extrusion. The measured values are recorded in pre-prepared tables.

Dilation of extrusions is a very important factor in the experimentation [17, 18]. It is a specific parameter existing especially in the compaction process. Dilatation of extrusions is a phenomenon during which the dimensions of the extrusion (diameter, length, and weight of the extrusion) change due to the action of internal and external parameters of the extrusion. In terms of processing and evaluating the experiments, it is not possible to work with the values obtained and measured immediately after extrusion. Dilatation directly affects the parameters of the extrusion. The dilatation process arises due to the release of pressure(s), and depending on the

Fig. 6.16 Sequence of extrusion throughout experimentation

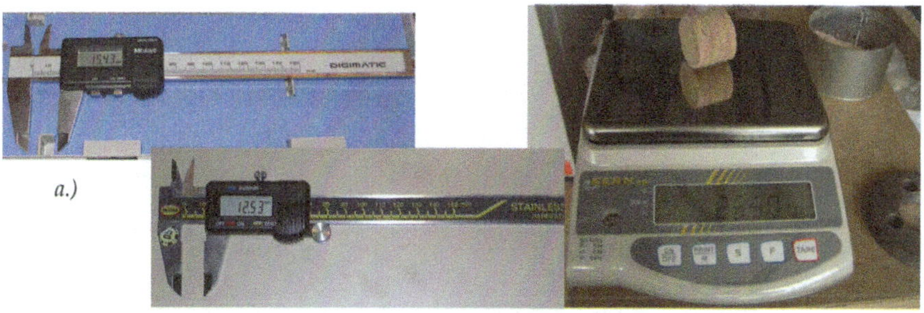

a.) *b.)* *c.)*

Fig. 6.17 measurement apparatus—(**a**) digital caliper 1 for diameter measurements; (**b**) digital caliper 2 for length measurements; (**c**) digital scale KERN

mutual interaction of the pressing temperature, the input moisture content, and fraction size of the pressed material. Throughout the experiments, we encounter two basic types of extrusion dilation [17, 18]:

- *Dilatation under the influence of a decrease in the density of the extrusion*—this usually happens if the diameter of the extrusion increases, the length of the extrusion increases, and the weight of the extrusion decreases (accompanying phenomenon maybe cracking, separation of material particles, or even crumbling of the extrusion).
- *Dilatation under the influence of an increase in density of the extrusion*—this usually happens if the diameter of the extrusion decreases, the length of the extrusion is shortened and the weight of the extrusion increases (accompanying phenomenon is shrinkage and hardening of the extrusion).

Both of these are, of course, significantly influenced by the type of pressed material, moisture content of the pressed material, the fraction size of the pressed material, the temperature during compaction, the pressure applied, and the extrusion time.

The stabilization time of the extrusion is the duration during which the extrusion dilates, the time during which the extrusion stabilizes [17, 18]. The stabilization time of a biomass extrusion is approximately 24 h depending on the material and pressing technology, but it may take longer. After compaction, the extrusion is placed in a stable environment (see Fig. 6.18). The diameter, length, and weight of the extrusion are measured multiple times, and its weight was evaluated over this period. If the weight of the extrusion changes by no more than 0.1% over 24 h, the condition of the extrusion is considered to be stable [19].

Considering the stabilization time, the dilatation of extrusions is of great importance especially with regard to the change in density. The density of extrusions is one of the primary mechanical indicators of quality, so it is determined in virtually every experiment. Density is important in terms of sufficient cohesion of the material particles within the extrusion. Therefore, it is important to consider dilation experiments and thus to process the data obtained on the extrusion only after stabilization. During compaction, lignin is released from the biomass, which acts as a

Fig. 6.18 Extrusions undergoing stabilization

binder. It forms a layer ("thin film") over the surface, which prevents absorption of moisture from the environment [20, 21]. If the extrusion is not formed under optimally defined conditions, the expansion of the extrusion can significantly affect the absorption rate of moisture from the environment. In industrial operations, failure to accept the effects of dilation can lead to disintegration of the extrusions. In production lines, the standard stabilization time is shortened by cooling the extrusions. This after treatment can significantly skew the results of experiments.

6.2 Results of the Experimental Research

All experiments were performed under stable laboratory conditions, using instruments and devices described within the previous chapter. All graphical dependences are derived from the average values of densities and other measured parameters of extrusions after their stabilization. The values obtained and calculated in this way can be considered as correctly determined. Some experiments required a change in material and technological parameters. To confirm/verify certain hypotheses, it was necessary in some cases to add different degrees of material and technological parameters used in the experimental research. Therefore, specific values at which the experiments were performed are not given here. The conditions and values of the experiment for such cases are described for each experiment separately. The determined graphical dependencies provide a detailed overview on the influence of the monitored parameters in the process of biomass compaction. Throughout the experimental research, the behavior of individually monitored parameters and the determination of their impact on the resulting quality of extrusions were performed as follows:

1. The influence of the pressing method
2. The influence of the pressing chamber length
3. The influence of the degree of pressing chamber taper (conicity)
4. Influence of the distribution and action of friction forces within the pressing chamber

6.2.1 Determining the Influence of the Pressing Method

The experiments were performed under laboratory conditions at ambient temperatures of approximately 23 °C. So that the pressing temperature did not affect the magnitude of the radial pressures and friction conditions, and additional heating of the chamber was not performed. The experiments used pine sawdust as input material with a material fraction size of 2 mm and moisture content of 11.5%. 7 extrusions were performed for each setting, and the compaction took place at three pressure levels—95 MPa, 127 MPa, and 159 MPa.

In this experiment, the aim was to determine the influence of different pressing methods on the final quality of extrusions and to obtain information on the contribution of radial pressures within the pressing chamber. Due to the conditions of the experimental equipment, the experiment was simplified (see Figs. 6.9 and 6.10). In "single-cycle" compaction, which represents a pressing cycle within a "closed" chamber (see Fig. 6.20), it can be seen from the density of extrusions that as the pressing pressure increases, the density of the extrusion increases (Fig. 6.19) and also the force required to push the extrusions out of the pressing chamber. After the stabilization period, there was a significant decrease in density of the extrusions, especially after the dilation process. This fact can be attributed to weaker bonding between materials due to the insufficient action of the pressing pressure. When pressing without additional heating, the pressing pressure is the primary effect on the resulting density of extrusions (Fig. 6.19).

An effort was also made to monitor and record the pressures required to move the extrusions through the chamber as a function of the applied axial compression pressures (Figs. 6.21, 6.22, and 6.23). The pressure required to move the extrusion and thus overcome the friction within the pressing chamber was also recorded for three levels of pressing pressure.

It can be seen that an increase in initial pressure (for compaction) increases the pressure required to overcome the friction between the extrusion and the walls of the pressing chamber. This phenomenon is expected and confirmed in more detail within the following Fig. 6.24. From the mentioned dependences, we can also determine the dependency of the compaction pressure on the position of the extrusion within the pressing chamber and on which material is pressed. We can see that when pressing in a 140-mm-long pressing chamber, the greatest friction between the extrusion and chamber walls was recorded at a distance of approximately 130 mm. This is, however, a question depending on the degree of compaction and thus the treatment and type of compressed material (moisture and fraction size).

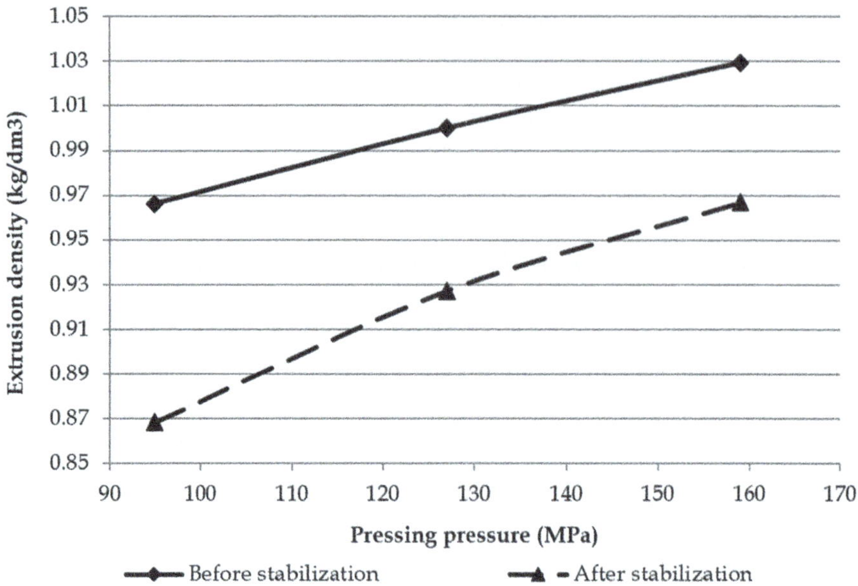

Fig. 6.19 Dependence of extrusion density on the pressing pressure within "single-cycle" compaction [22, 23]

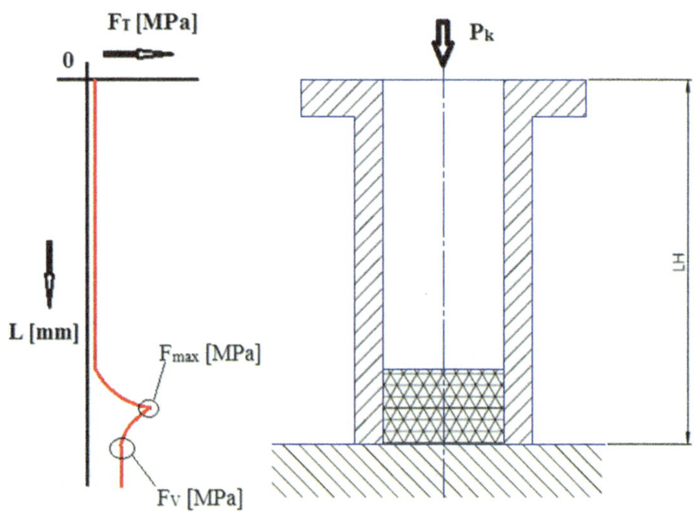

Fig. 6.20 Graphical representation of the behavior of acting forces within the pressing chamber in "single-cycle" compaction [8]

In addition to experiments, where only one extrusion is pressed through the nozzle, measurements were also performed in "multi-cycle" pressing. The experiments were performed for double and triple cycle pressing (Fig. 6.25) with an "open"

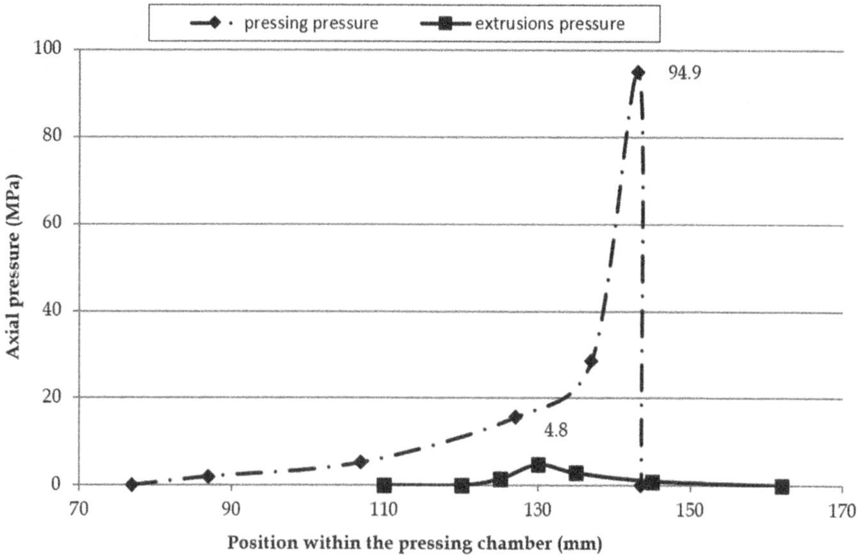

Fig. 6.21 Dependence of pressure required for compaction (95 MPa) and pressure required to eject the extrusion with respect to its position in the pressing chamber

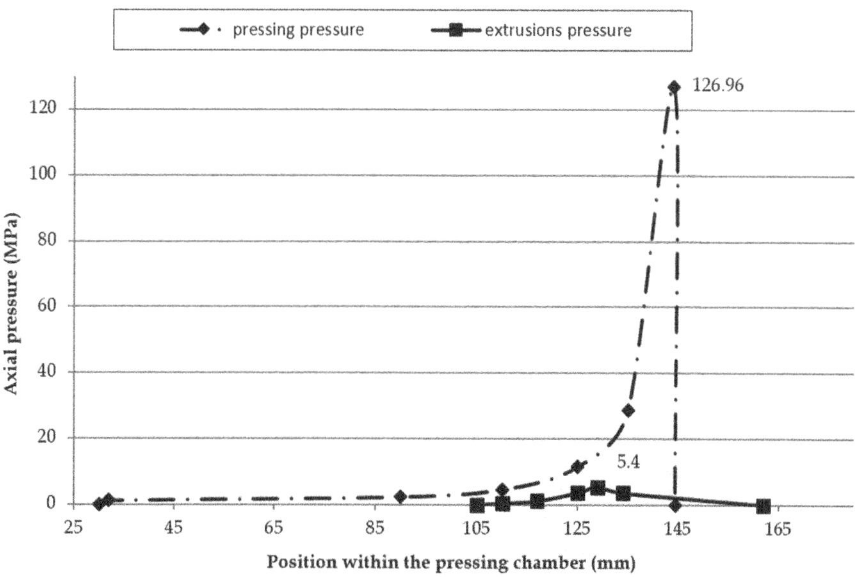

Fig. 6.22 Dependence of pressure required for compaction (127 MPa) and pressure required to eject the extrusion with respect to its position in the pressing chamber

pressing chamber. After ejecting an extrusion from the pressing chamber, they were weighed and measured (length, diameter). After extrusion of an entire column from the chamber, individual extrusions were separated and measured separately. All

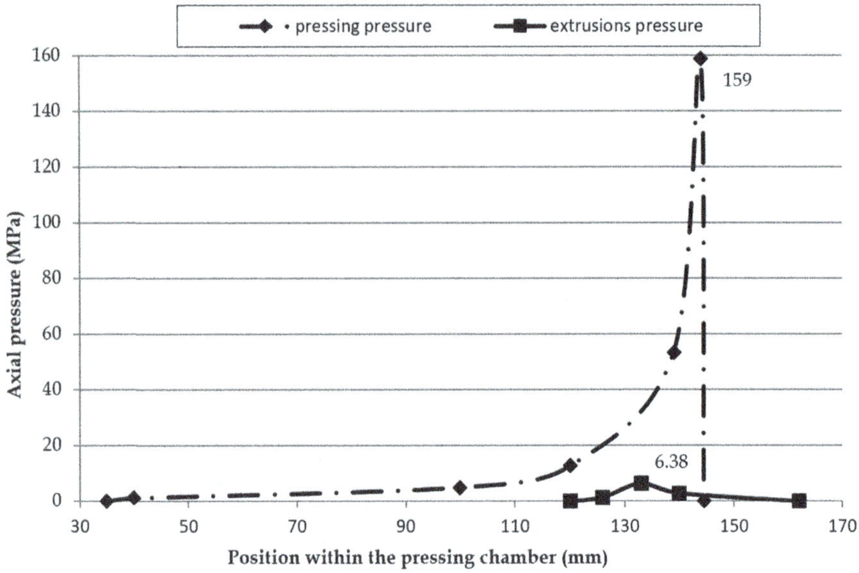

Fig. 6.23 Dependence of pressure required for compaction (159 MPa) and pressure required to eject the extrusion with respect to its position in the pressing chamber

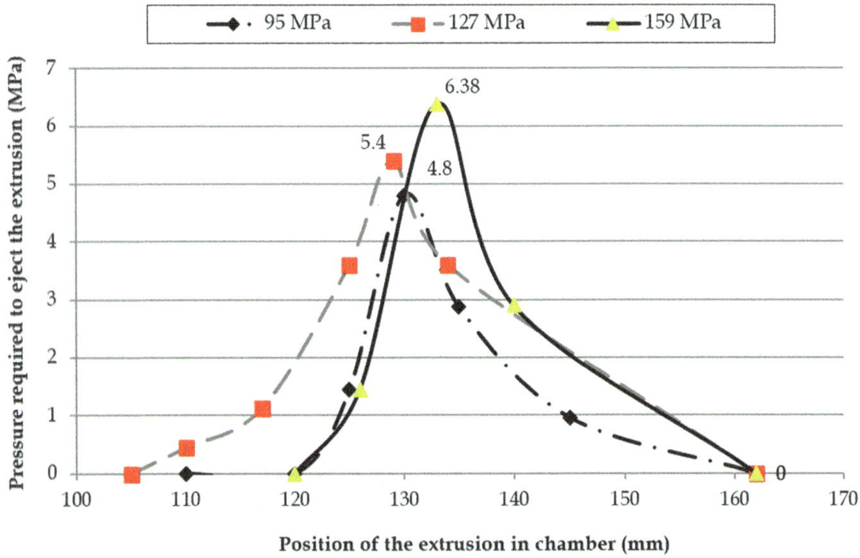

Fig. 6.24 Behavior of pressure required to eject the extrusions with respect to the pressing pressure

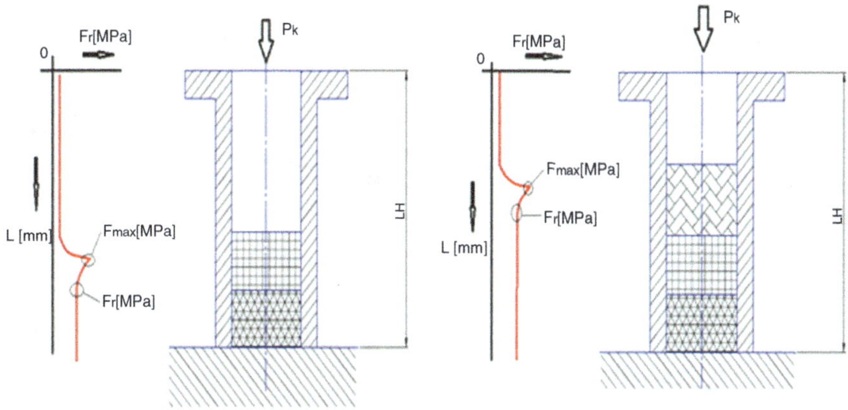

Fig. 6.25 Graphical representation of the behavior of forces acting within the pressing chamber during "double" and "triple" cycle pressing [8]

graphical dependences were compiled from average values at individual pressure levels.

The following Fig. 6.26 shows a comparison of densities obtained after single, double, and triple cycles of pressing with sawdust as the input material. It can be seen that with an increase in the number of pressing cycles, the time during which the extrusion is subject to the actions from the compaction pressure within the pressing chamber, extrusions also increased in density.

These results clearly show that it is possible to obtain extrusions with a higher density, and thus a higher final quality, after compaction in an "open" pressing chamber. This is due to the fact that during "multiple" compaction cycles, the pressure in the chamber acts on the extrusion for a longer duration. Furthermore, the extrusion is under the influence of these pressure conditions long enough that the internal radial pressures do not rupture the extrusion after ejecting it from the chamber.

Figure 6.27 shows the dependence of extrusion densities on the pressing order within the pressing chamber. The first three extrusions, at all pressure levels, display only slight fluctuations in density values. The density values differ very slightly for the first three extrusions. Extrusion no. 4 does show a decrease in density. These extrusions were not considered in the evaluation of the previous experiment. The depicted graphical dependence is given for documentation of the pressures acting within the pressing chamber. Extrusions no. 4 was subject to pressing pressure only once and its duration spent within the pressing chamber was the shortest, which means that there is insufficient duration for the equalization of internal stresses. Figure 6.27 shows that the density of the extrusions increases as the length of the compacted column increases and as the dwell time within the pressing chamber increases.

The pressure profiles required for compaction of the extrusions as a function of the axial compaction pressures acting separately in "double" and "triple" cycle

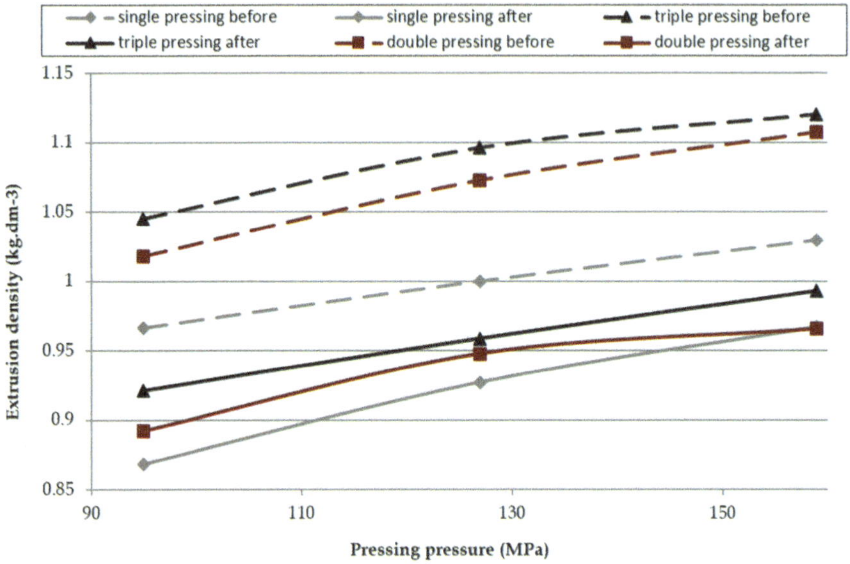

Fig. 6.26 Comparison of the dependencies of extrusion densities on the compaction pressure in single, double, and triple compaction

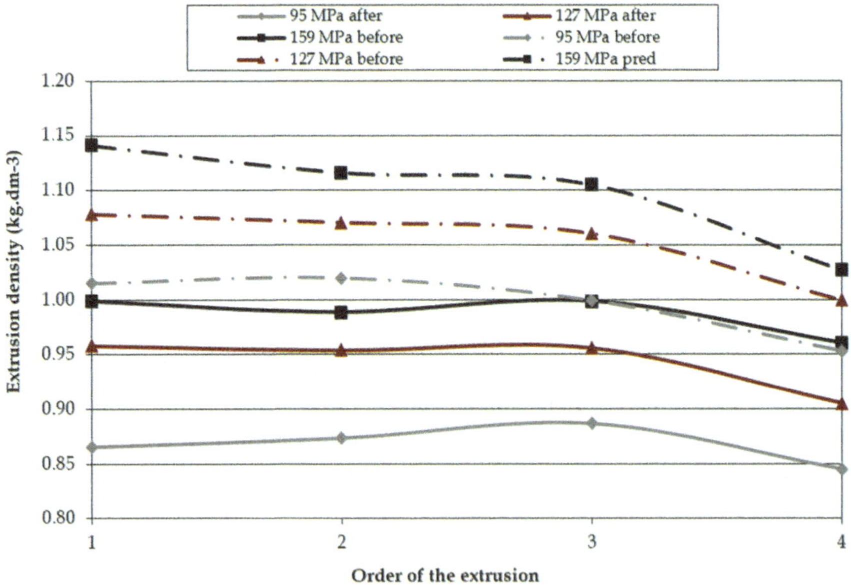

Fig. 6.27 Dependence of extrusion density on the order of the extrusion within the pressing chamber and on the applied pressing pressure

pressing are shown in the following Figs. 6.28 and 6.29. For clarity, only the behaviors of "double" and "triple" cycle pressing are shown for a compaction pressure of 95 MPa. However, in comparison with the pressure profiles required for compaction

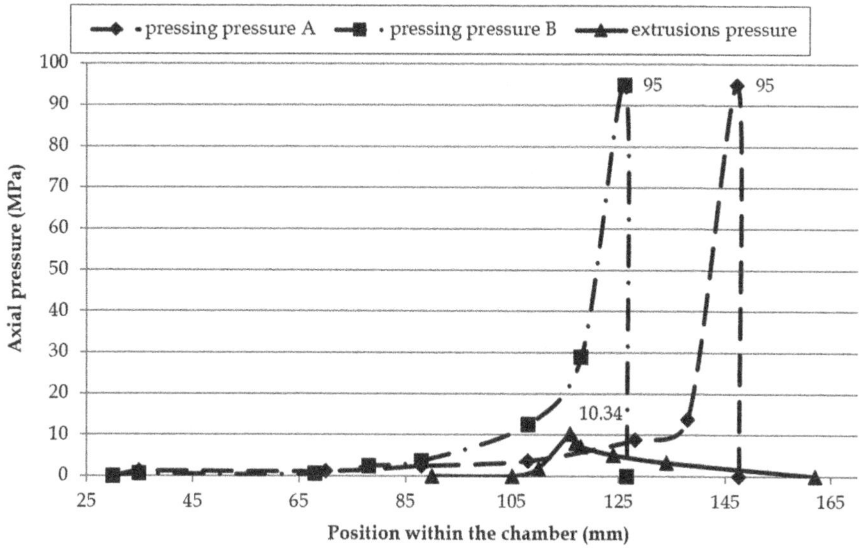

Fig. 6.28 Dependence of the pressure required to eject the extrusion in "double" pressing with respect to its position within the pressing chamber under the influence of a compaction pressure of max. 95 MPa

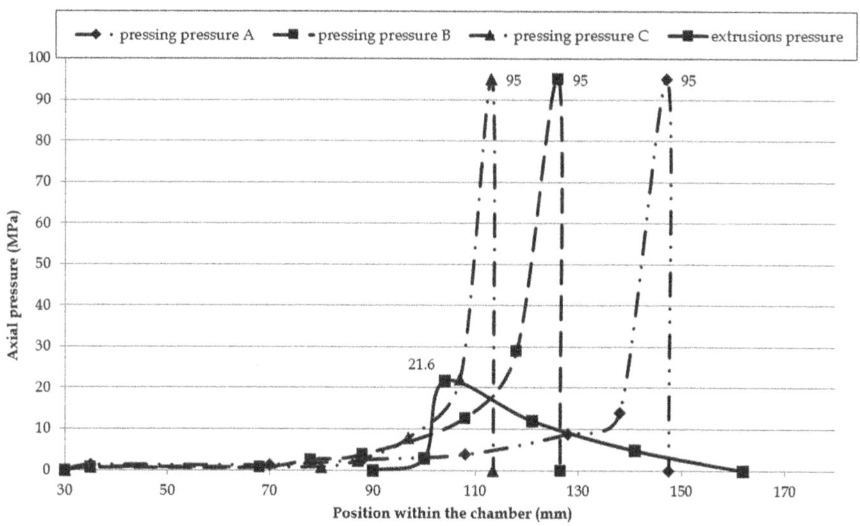

Fig. 6.29 Dependence of the pressure required to eject the extrusion in "triple" pressing with respect to its position within the pressing chamber under the influence of a compaction pressure of max. 95 MPa

(Figs. 6.30 and 6.31), the values recorded during "double" and "triple" pressing at compaction pressures of 127 and 159 MPa are also given.

From these dependences, it can be seen that by changing the pressing method, it is possible to significantly influence the resulting density of the extrusions and also

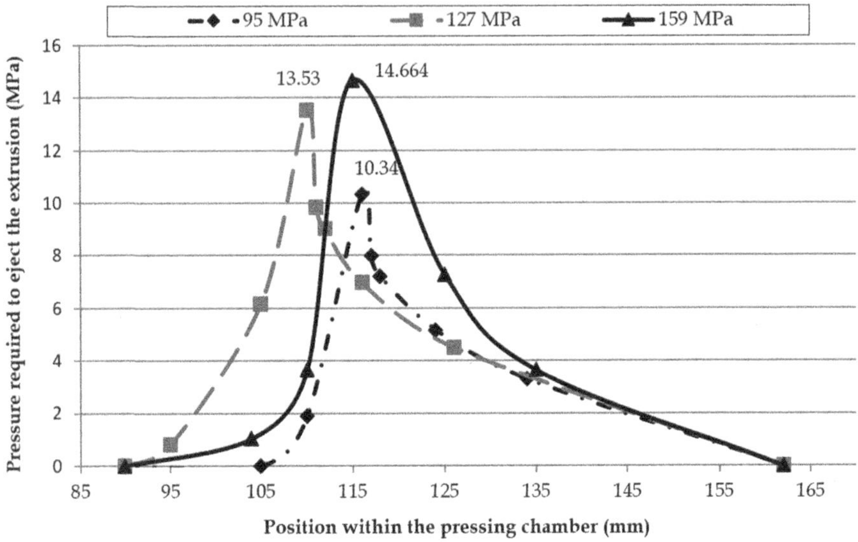

Fig. 6.30 Behavior of the ejection pressure with respect to the position of the extrusion within the pressing chamber in "double" pressing

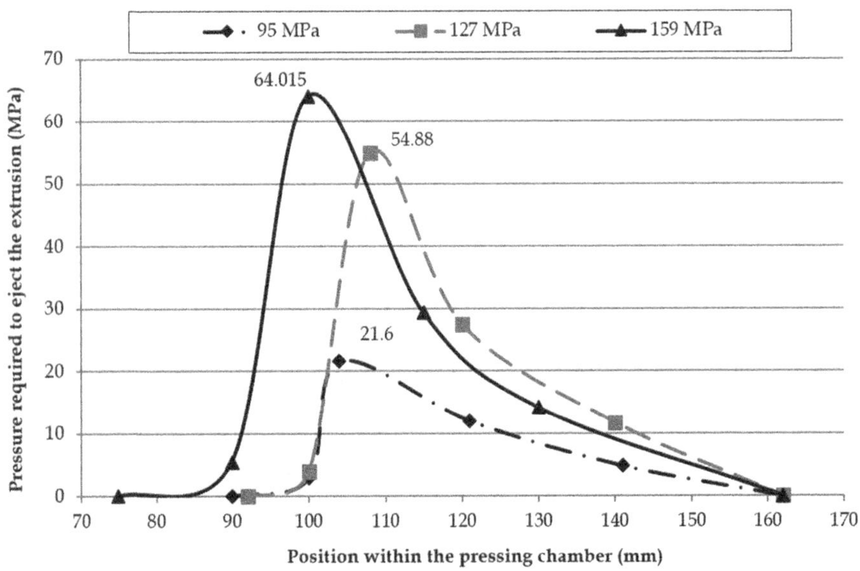

Fig. 6.31 Behavior of the ejection pressure with respect to the position of the extrusion within the pressing chamber in "triple" pressing

the friction conditions within the pressing chamber. A comparison of the pressures required to extrude material over a 140-mm-long pressing chamber with respect to the pressing method is given in Table 6.2. The recorded pressure required for

Table 6.2 Table of calculated extrusion density and corresponding pressures

	95 MPa		127 MPa		159 MPa	
Pressing	P_V (MPa)	ρ (kg.dm^{-3})	P_V (MPa)	ρ (kg.dm^{-3})	P_V (MPa)	ρ (kg.dm^{-3})
1x	4.8	0.868	5.4	0.927	6.38	0.965
2x	10.34	0.892	13.53	0.947	14.664	0.967
3x	21.6	0.921	54.88	0.959	64.015	0.993

Legend: P_V required ejection pressure, ρ extrusion density

extrusion in "single" pressing is approximately 50% lower than in "double" pressing, at all levels of compaction pressure. Subsequently, in "triple" pressing, the pressure required for extrusion increases exponentially when compared to "double" pressing. The higher pressures necessary for pushing the extrusion through the pressing chamber result in more suitable pressure conditions, resulting in higher density extrusions. In the previous Figs. 6.30 and 6.31, we can also observe different positions of the extrusion within the pressing chamber, where the greatest friction is recorded depending on the compaction pressures and the pressing method.

In principle, it can be stated that the method of pressing affects the resulting quality of the extrusion and also the distribution of pressures within the pressing chamber. This result therefore shows better efficiency in the production of pressing mills, which work on the principle of an "open" pressing chamber, in comparison with the a "closed" pressing chamber. This result is verified in Fig. 6.32, which compares the values of extrusion densities obtained in "single" pressing and in "multiple" (continuous) pressing methods. This method was achieved on an experimental pressing stand, after releasing the back pressure plug. After refilling the chamber several times with the pressed material and compacting, a gradually compacted column of material is extruded out of the pressing chamber (Fig. 6.33). This perfectly simulated the pressing method, which is mostly used in mechanical briquette presses and in pellet mills.

6.2.2 *Defining the Effects of the Chamber Length*

The aim of the experiment described hereafter was to determine the effect of the pressing chamber length. As described in the previous chapter, variable length pressing chambers were designed and prepared for this purpose. All experiments were performed under laboratory conditions at an ambient temperature of approximately 21 °C. The results were evaluated in terms of the final quality of the extrusions, represented by their density. Graphical dependences are derived from the calculated values of densities after stabilization. In determining the effect of chamber length, it was appropriate and necessary to carry out experiments in both "single" and "multiple" pressing cycles. The influence of the length of the pressing chamber with and without additional heat was also determined. The effect of temperature on the pressing process has been experimentally proven and verified in the past.

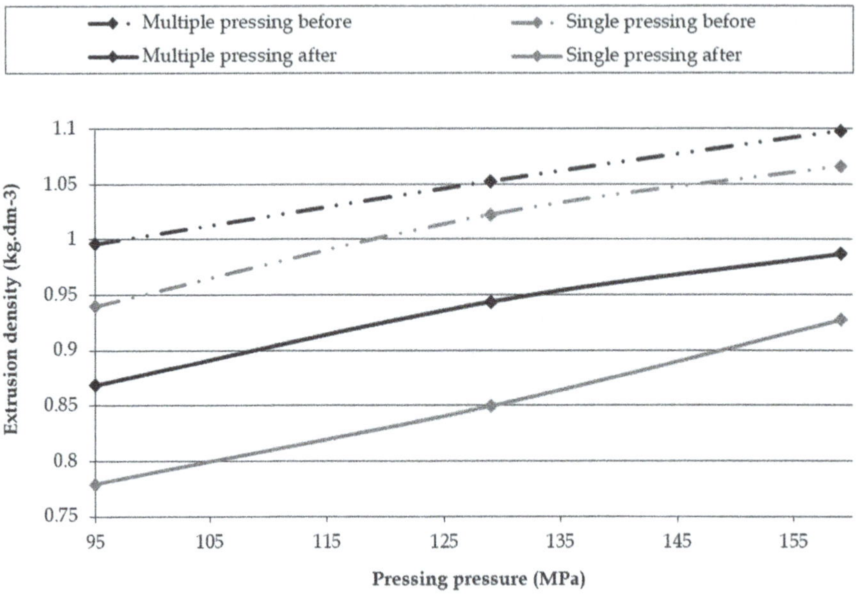

Fig. 6.32 Dependence of the extrusion density on the pressing pressure between "single" and "multiple" pressing(s) over a chamber length of 140 mm

Fig. 6.33 Continuous extrusion being pushed out of the pressing chamber

In the following Figs. 6.34, 6.35, 6.36, 6.37 and 6.38, it is possible to see the dependences of extrusion densities on the compaction pressure, without any additional heating. In general, it can be said that the densities of extrusions were not very

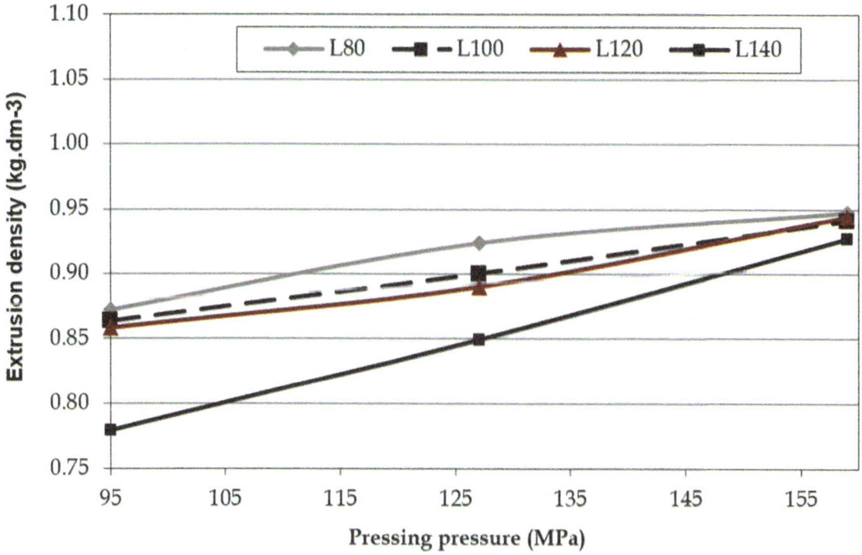

Fig. 6.34 Dependence of extrusion density on the pressing pressure in "single" pressing with various chamber lengths

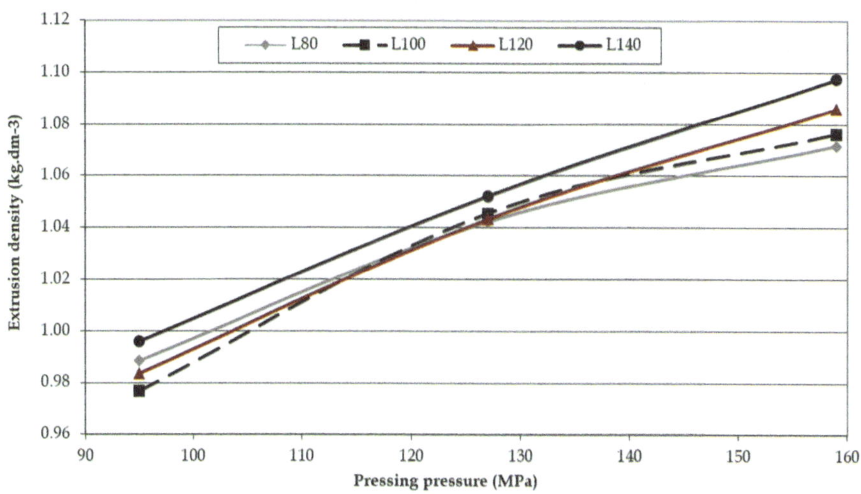

Fig. 6.35 Dependence of the extrusion density on the pressing pressure in "multiple" pressings in chambers with different lengths

different, but the results do show a difference between "single" and "multiple" pressing cycles.

In the case of "single" pressing, the extrusions with higher density were pressed in a pressing chamber 80 mm long. In the case of "multiple" pressing, the results are opposite, and extrusions with the highest quality were obtained in a pressing

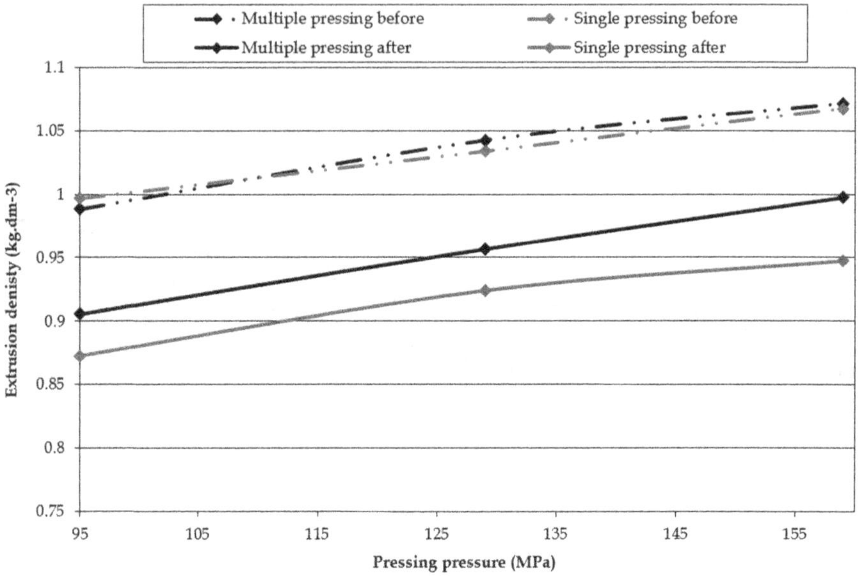

Fig. 6.36 Dependence of the extrusion density on the pressing pressure in "single" and "multiple" pressing(s) within a chamber length of 80 mm

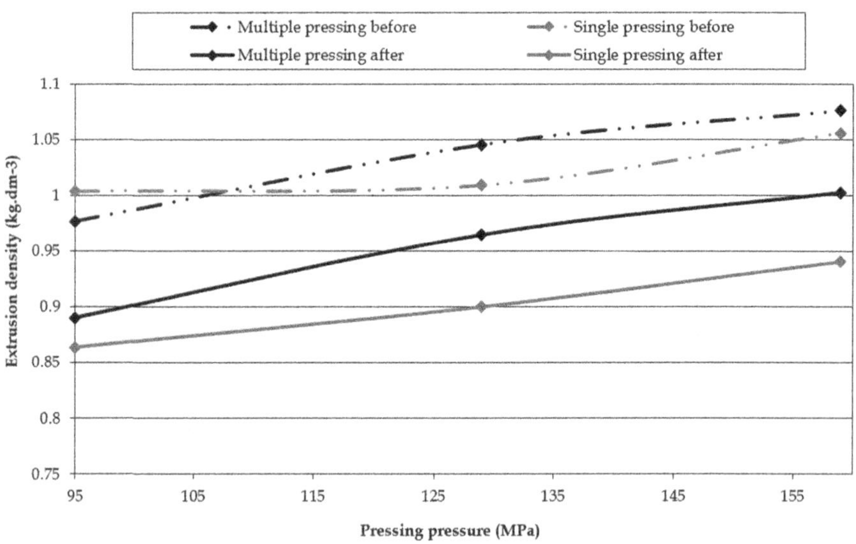

Fig. 6.37 Dependence of extrusion density on the pressing pressure in "single" and "multiple" pressing(s) with a chamber length of 100 mm

chamber of 140 mm long. Repeated application of the compaction pressure on the extrusions throughout "multiple" pressing has a positive effect on their density. The difference between the aforementioned pressing methods is also apparent with different length pressing chambers. This is related to the different volume of the

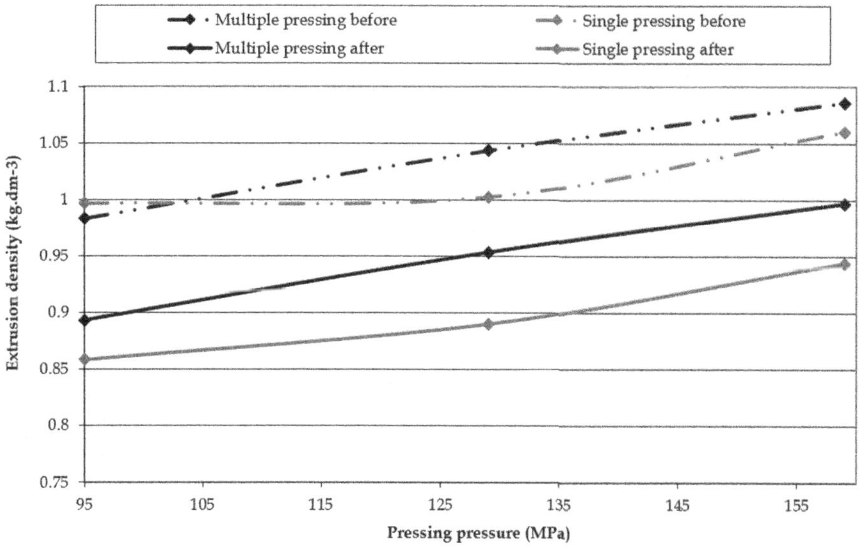

Fig. 6.38 Dependence of extrusion density on the pressing pressure in "single" and "multiple" pressing(s) with a chamber length of 120 mm

pressing chamber and thus the amount of pressed material with which the chamber can be filled. When pressing with constant compaction pressures, a smaller amount of input material is compressed better. This is apparent in "single" pressings.

However, by comparing the extrusion densities of "single" and "multiple" pressing(s) for each length of the press chamber, we see that higher densities were achieved after dilation in "multiple" pressing for each chamber length. Figure 6.32 shows a comparison when pressing in a 140-mm chamber. Figures 6.36, 6.37, and 6.38 show comparisons in chamber lengths of 80, 100, and 120 mm.

It can be seen that as the length of the pressing chamber increases, so do the differences between the densities of extrusions pressed by "single" and "multiple" pressing(s). The results presented in this section confirm that higher extrusion densities can be achieved by "multiple" pressings. It is also true that as the length of the pressing chamber increases in "multiple" pressing, the density of the extrusions also increases (see Table 6.3).

In the following part, the influence of pressing chamber length with additional heating was experimentally determined. The experiments were performed at 3 levels of pressing temperature, namely 23, 85, and 120 °C (see Figs. 6.39, 6.40, 6.41, 6.42 and 6.43). The influence of temperature on the pressing process has been experimentally proven and verified in the past, and the results obtained experimentally serve to confirm these findings.

The effect of increasing the temperature during compaction is evident over all lengths of the pressing chamber. This is mainly due to the temperature causing the lignin in the cellular structures of the pressed material to plasticize [24]. When compacting without additional heat, the extrusion undergoes significant expansion as it is ejected from the pressing chamber, causing it to crack and disintegrate. On the

Table 6.3 Comparison of calculated extrusion densities for different press settings

	Single pressing → ρ (kg.dm^{-3})				Multiple pressing → ρ (kg.dm^{-3})			
Length L_k	80 mm	100 mm	120 mm	140 mm	80 mm	100 mm	120 mm	140 mm
Pressing pressure								
95 MPa	0.872	0.863	0.858	0.779	0.988	0.977	0.984	0.996
127 MPa	0.924	0.899	0.890	0.849	1.042	1.045	1.043	1.052
159 MPa	0.947	0.940	0.943	0.927	1.072	1.076	1.086	1.096

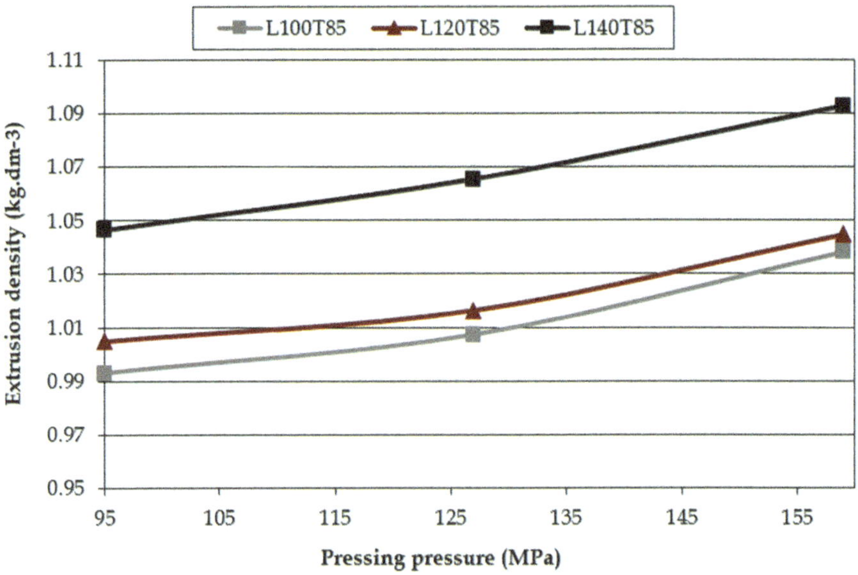

Fig. 6.39 Dependence of extrusion density on the pressing pressure at 85 °C with varying chamber length

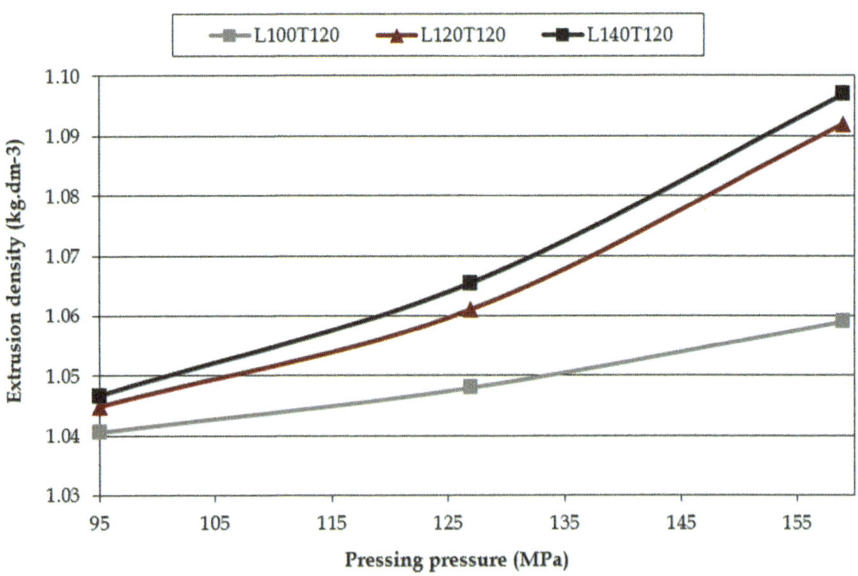

Fig. 6.40 Dependence of extrusion density on the pressing pressure at 120 °C with varying chamber length

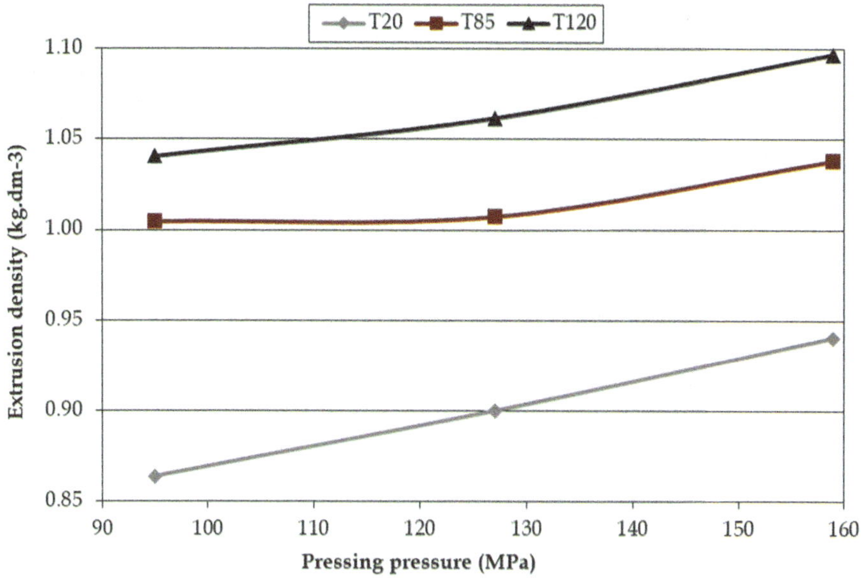

Fig. 6.41 Dependence of the extrusion density on the pressing pressure over a chamber length of 100 mm and under the influence of different pressing temperatures

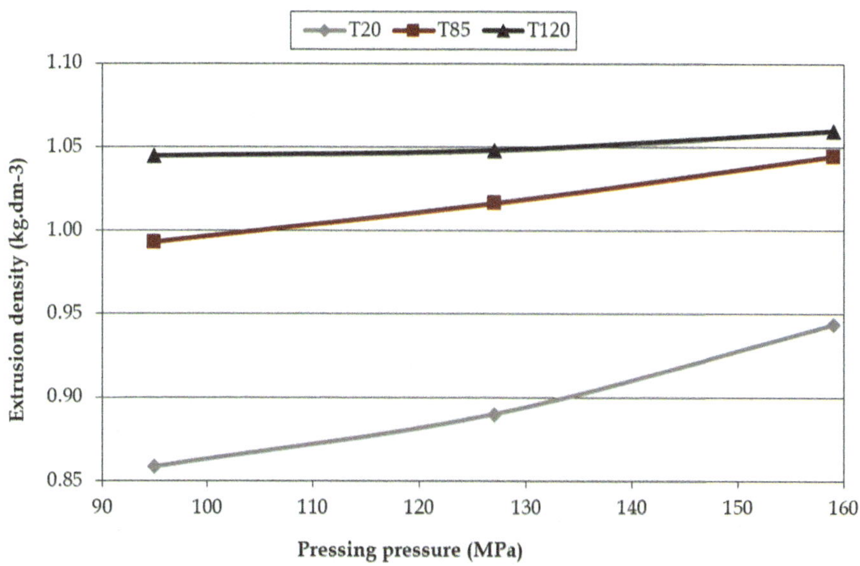

Fig. 6.42 Dependence of the extrusion density on the pressing pressure over a chamber length of 120 mm and under the influence of different pressing temperatures

contrary, especially when pressing at higher temperatures, the resulting surface was smooth and more compact, which allows for better handling of the extrusions.

This is also evidenced by the following graphical dependencies, which show the increase in extrusion density as a function of the increase in compaction pressure

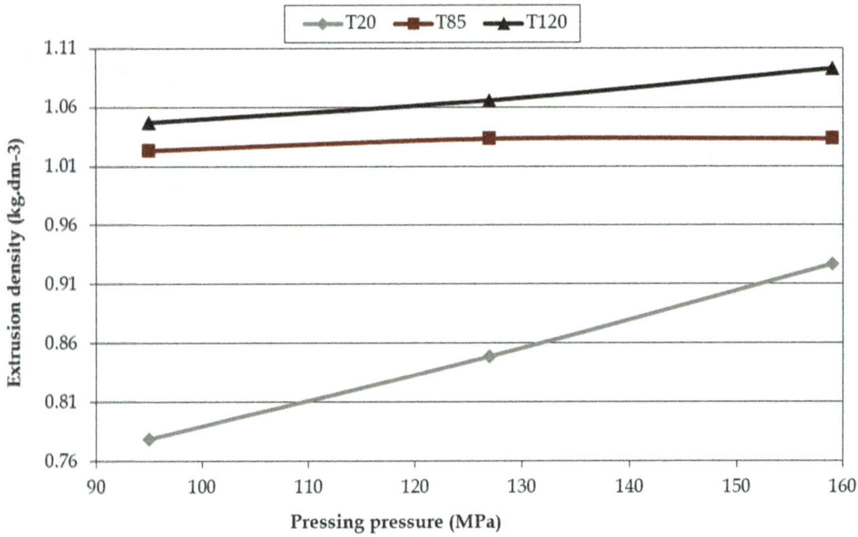

Fig. 6.43 Dependence of the extrusion density on the pressing pressure over a chamber length of 140 mm and under the influence of different pressing temperatures

and temperature. These graphical dependencies are created for each monitored length of the pressing chamber separately. However, we do not compare the densities when pressing in an 80-mm pressing chamber between the above graphs. This is because this pressing chamber could not be sufficiently heated due to its short length and due to the limited dimensions of the pressing stand. With this pressing chamber, it would not be possible to regulate the pressing temperature. Despite this, it has been confirmed that a suitable combination of chamber length, compaction pressure, and pressing temperature can achieve a higher extrusion density. Thus, these parameters can be considered in the optimization of the pressing process.

It should be noted that the graphical dependencies within Figs. 6.41, 6.42, and 6.43 were derived from the results of experiments performed in a "single" pressing method. Therefore, it is logical that with an increase in the length of the pressing chamber and the pressing temperature, no significant increase in the density of the extrusion is observed. The reason is the larger volume of material that must be heated in a longer pressing chamber and the different pressing distance of the extruded "plug" within the chamber.

Furthermore, the behavior of pressures necessary to eject the extrusions with respect to axial compression pressure was also observed. However, this time the chamber length and pressing temperature were also considered for "single" and "multiple" pressing(s). Figure 6.24 shows the behavior for "single" pressing over a chamber length of 140 mm. Figure 6.30 shows the behavior for "double" pressing and Fig. 6.31 shows the behavior for "triple" pressing, both over a chamber length of 140 mm. Since two more press chamber lengths (100 and 120 mm) were available for 3 pressing methods, another 6 graphical plots were created to show the

Table 6.4 Table of recorded pressures required to push the extrusion "P_V" (MPa) through the pressing chamber

	Length "L_k"			
Pressing pressure	Pressing method	100 mm	120 mm	140 mm
95 MPa	1x	4.13	4.8	4.8
	2x	8.5	9.215	10.34
	3x	12.1	34.5	21.6
127 MPa	1x	5.4	5.4	5.4
	2x	9.6	12	13.53
	3x	13.23	37.23	54.88
159 MPa	1x	6.72	6.38	6.38
	2x	13.52	14.21	14.664
	3x	24.255	38	64.015

behavior of the pressing pressures. For clarity, the results are summarized in Table 6.4, which compares the maximum recorded pressure required to eject the extrusion with respect to the chamber length, the value of the axial compaction pressure, and the pressing method used.

Based on the pressures recorded in the table above, it can be stated that in the case of "single" pressing, the change in chamber length did not significantly affect the pressure profiles when pushing the extrusions out of the chamber. This is logical and expected, since the volume of pressed material with which the pressing chambers were filled is approximately the same (depending on the length of the pressing chamber). In the case of "double" and "triple" pressings, different chamber lengths required different pressure profiles to move the extrusion throughout the chamber. As the lengths of the pressing chamber increased, the differences increased, and as the length of the pressing chamber increased, so did the pressure required to push the extrusion out of the chamber. This is due to the different distribution of pressures on the already compacted extrusions "plugs" within the pressing chamber, and thus the different friction conditions with which they coincide.

6.2.3 *Defining the Effects of Pressing Chamber Taper "Conicity"*

Another part of the experimental research was to determine the effect of the pressing chamber conicity (chamber wall taper). In order to investigate this effect, 3 conical pressing chambers with varying degrees of taper were designed and manufactured for experimentation. All experiments were performed under laboratory conditions at an ambient temperature of approximately 23 °C. The results were evaluated in terms of the final quality of extrusions, represented mainly by their density. Graphical dependencies are created from calculated values of densities after stabilization. In this experiment, only the "multiple" pressing method was considered since using the "single" pressing method would be impossible to extract the compacted material. In carrying out the experiments investigating the effect of pressing chamber conicity, the additional effects of pressing temperature proved to be a necessary condition for successful extrusion. As shown in Table 6.5 below, without the application of an additional pressing temperature (at an ambient temperature of 23 °C), it was not possible to eject the extrusion from the chamber. The limiting factor was an insufficient axial pressure of the hydraulic press. As mentioned in the previous chapter, due to the construction of the experimental pressing stand, the hydraulic press makes it possible to derive a maximum axial pressure of 318 MPa. Even when applying the maximum possible axial pressure that the hydraulic press could generate, it was not possible to extrude the compacted material at certain experimental settings.

Based on the results obtained from the experiments, it can be argued that a change in conicity of the pressing chamber significantly affects the resulting quality of extrusions. Also, when comparing the results from pressing in a cylindrical pressing chamber with pressing in a conical pressing chamber, under the same experimental conditions, the extrusions are of a significantly higher quality. The extrusions even surpassed the minimum density value of 1.14 ($kg.dm^{-3}$) defined in standards. The pressed material is sufficiently compacted due to the use of a tapered pressing chamber. Due to the distribution of the pressures acting within the conical chamber, higher values of radial pressures are induced, which influence the occurrence of friction. From a technological point of view, the compaction process has changed from discontinuous to continuous. The individually compressed extrusions "plugs" were perfectly interconnected, resulting in a pressing process, which could be repeated cyclically (filling–pressing). The process compressed the input material, which resulted in a subsequent displacement of the entire column compacted within the pressing chamber followed by its extrusion out of the pressing chamber. Figure 6.44 shows the extrusion, which exhibits signs of high quality, in terms of density. The extrusions pressed as described did not show any signs of cracking or disintegration, and were uniformly compressed over its entire length.

The boundary shape of the extrusions actually shows the action and distribution of pressures throughout the cross section of the pressing chamber. They prove that the extrusions are more flexible closer to the pressing axis when compared to a surface where radial pressures and friction are acting (see Fig. 6.45).

Of interest are also the recorded pressures required for extrusion during pressing in the conical pressing chamber, as well as the behavior of the pressing and subsequent extrusion. Initially, after releasing the back pressure plug, the extrusion is

Table 6.5 Table of calculated extrusion densities "ρ" ($kg.dm^{-3}$) [6]

Pressing chamber taper angle	$T_1 = 23$ °C	$T_2 = 85$ °C	$T_3 = 120$ °C
$\alpha = 0°$	0.852	1.152	1.207
$\alpha = 1°$	–	1.206	1.236
$\alpha = 2°$	–	1.224	1.241
$\alpha = 3°$	–	–	–

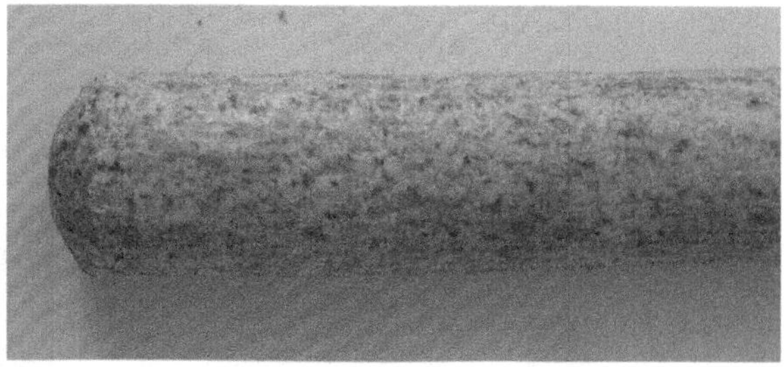

Fig. 6.44 Continuous extrusion

Fig. 6.45 Graphical representation of the behavior of pressures throughout the cross section of the pressing chamber

ejected from the pressing chamber as a non-compact and disintegrating. However, once a sufficient back pressure was established due to the taper in the pressing chamber walls, the counterpressure created by the length of the extruded column and its resulting friction had to be overcome. This pressure is referred to as the required "puncture" pressure. It is the pressure necessary to overcome the radial and frictional effects in order for the extrusion to be ejected from the chamber. Utilizing the instruments of the hydraulic press, it was possible to record and determine the magnitude of this pressure.

Overcoming this maximum pressure was accompanied by a notable sound effect—a significant cracking, which was the material overcoming friction and releasing tension. After overcoming this back pressure, a lower back pressure was recorded for each subsequent filling and pressing. Here, the freshly filled material was pressed and the compacted column is displaced. For this reason, the pressure was referred to as the "displacement" pressure. The pressure required to displace the compressed column within the pressing chamber and the pressure required to eject the extrusion from the chamber decreased depending on the length of the compacted column within the pressing chamber.

Based on the pressures recorded (Fig. 6.46), it is clear that the taper angle (chamber conicity) affects not only the resulting density of the extrusions, but also the distribution and action of the pressures within the pressing chamber. It can be seen that when pressing in a conical chamber with a 2° chamber taper, higher pressing pressures are required to puncture and displace the extrusions. The effect of chamber temperature on these pressures is also clearly visible. As described above, when

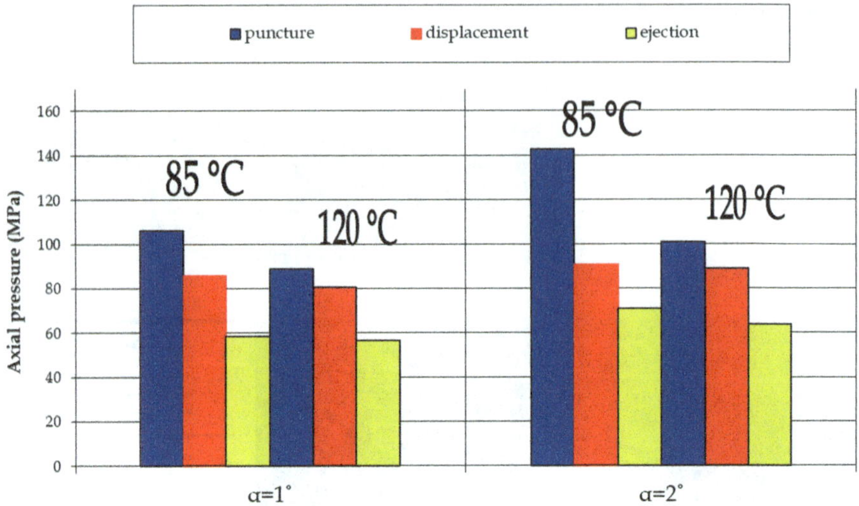

Fig. 6.46 Comparison of pressures required to puncture, displace, and eject an extrusion [6]

the temperature increases, the lignin within the cellular structure of the pressed material plasticizes. During pressing, this lignin also reaches the surface of the extrusions, i.e., between the pressed material and the walls of the chamber, and has a positive effect on the friction conditions. Therefore, depending on the types of biomass material to be compressed, it is desirable to create conditions with an increased operating temperature during compaction. This temperature refines the pressing process and also affects the final quality of the extrusions. In practice, some operators of production mills attempt to achieve this by additional preheating of the raw material, before they enter the pressing chamber (see Figs. 6.47, 6.48, and 6.49). In most cases, however, the increase in temperature is caused by friction within the pressing chamber, or by the use of an additional heating element.

As mentioned above, it was not possible to press material through a conical chamber with 3° of taper. The degree of the taper angle within the pressing chamber resulted in pressure conditions that were insurmountable for the construction of the hydraulic press. During these experiments, the pressing chamber would become clogged and the axial pressures could not overcome the puncture pressure to displace the compressed column out of the pressing chamber even when using additional heating. Therefore, the remnants of the pressed column had to be removed manually from the pressing chamber. Figure 6.50 shows the shapes of the manually removed remnants of the compressed column, which represents the so-called negative of the pressing chamber. It can be noticed that the surfaces of the extrusions are smooth and glossy indicating perfect bonding of the material particles and the action of high pressures.

The previous Figs. 6.51, 6.52, and 6.53 show the pressure profiles during compaction with conical pressing chambers at different chamber temperatures. These curves were obtained directly from the instrumentation of the hydraulic press. We

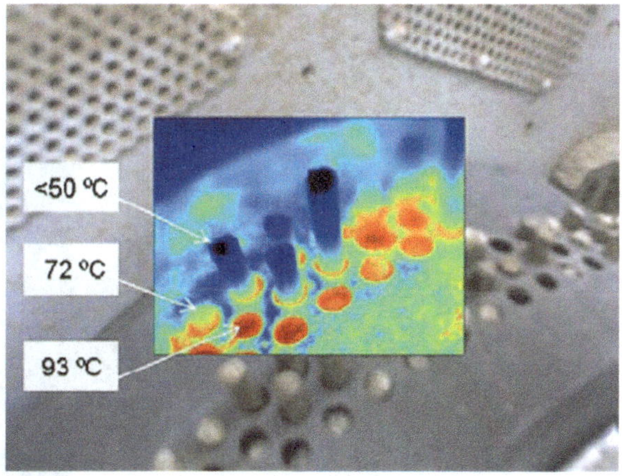

Fig. 6.47 Thermal image of a pellet mill die [25]

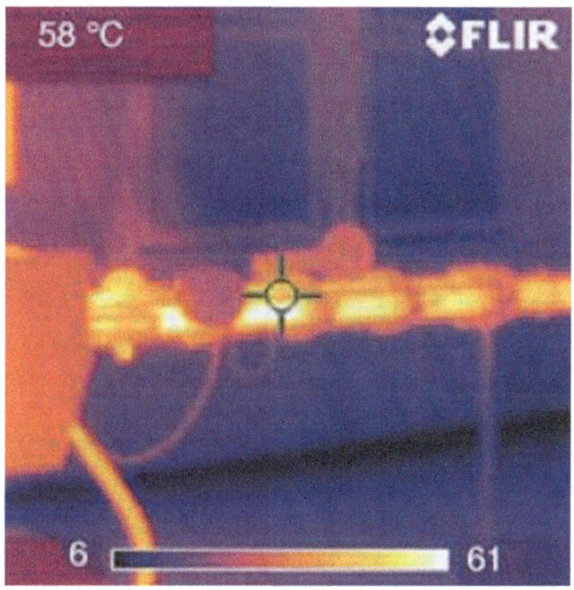

Fig. 6.48 Thermal image of the cooling channel of a briquette pressing chamber [26]

notice that the maximum pressures to be overcome range from 80 to 90 MPa, which is significantly higher compared to the pressures required for extrusion utilizing cylindrical chambers (see Table 6.4). It should be noted that these values are already in the presence of additional heat, which significantly softens the friction conditions within the pressing chamber.

Since the pressing temperature seemed irreplaceable in this process, it is interesting to show the quality of the extrusions obtained in conjunction with the compaction pressure and temperature (Fig. 6.54). In the production of solid biofuels, the ratio of temperature to compaction pressure can be expressed in the financial costs of biofuel production.

Fig. 6.49 Example of an additional heating element installed on a briquette press

Fig. 6.50 Remnants of compressed columns within the pressing chamber (pressing chamber negative)

The numerical expression of such economic costs depends on the specific values of the attachments used and the kinematics of the pressing machine. However, for a simple graphical representation of the dependence of extrusion density with respect to the temperature (T) to pressing (compression) pressure (p) ratio, it is not necessary to express the economic costs numerically. The ratio of pressing temperature to compaction pressure is sufficient. It can be seen in Fig. 6.54 that the selected (desired) compact density can be achieved by various combinations of the investigated conical chambers, the compression pressure, and the chamber temperature. The differences between the extrusion densities achieved are not large, and all values meet the requirements of standards. Therefore, it is possible to choose a setting of the technological conditions (compaction pressure, chamber temperature) from

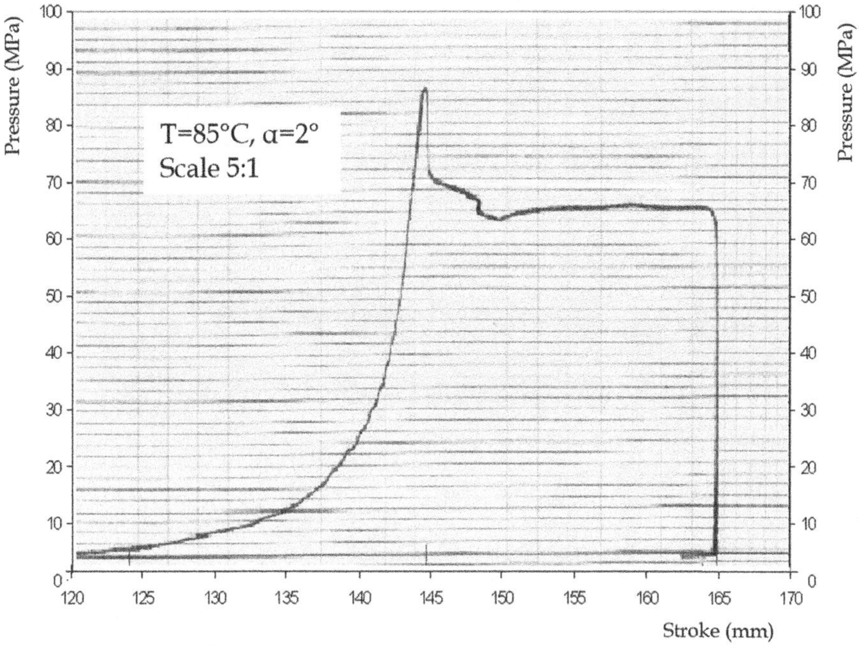

Fig. 6.51 Behavior of the pressing pressure when pressing through a conical chamber with 2° of taper at 85 °C chamber temperature [6]

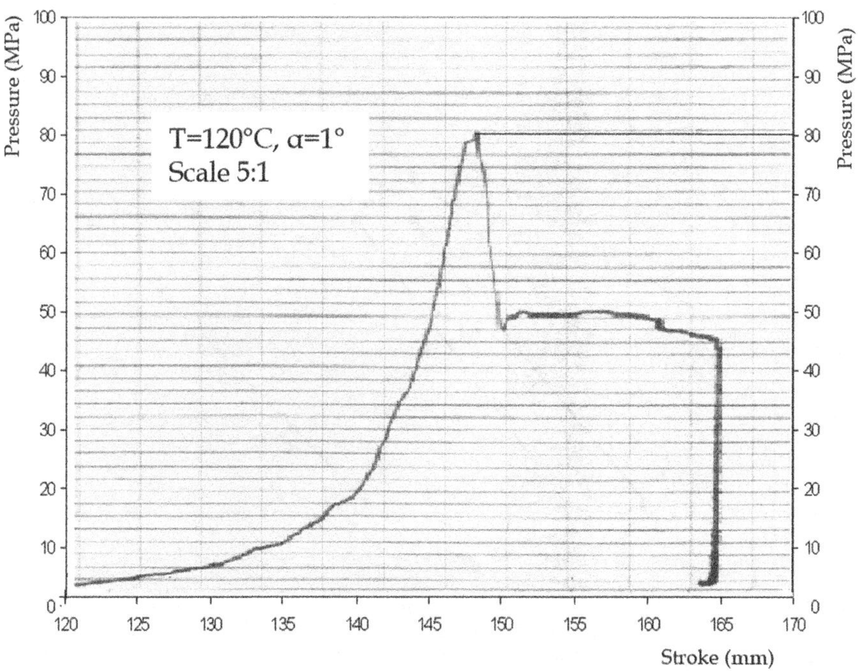

Fig. 6.52 Behavior of the pressing pressure when pressing through a conical chamber with 1° of taper at 120 °C chamber temperature [6]

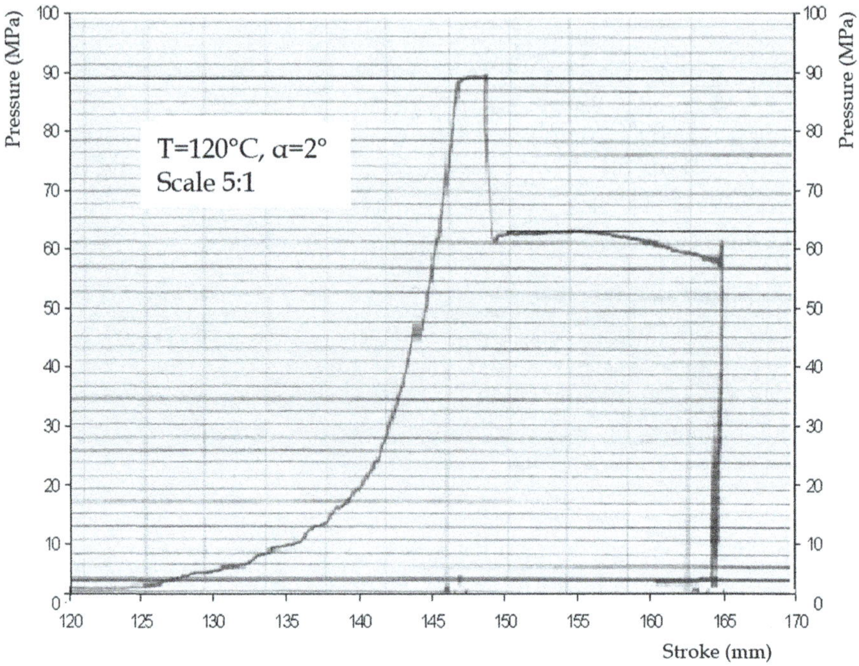

Fig. 6.53 Behavior of the pressing pressure when pressing through a conical chamber with 2° of taper at 120 °C chamber temperature [6]

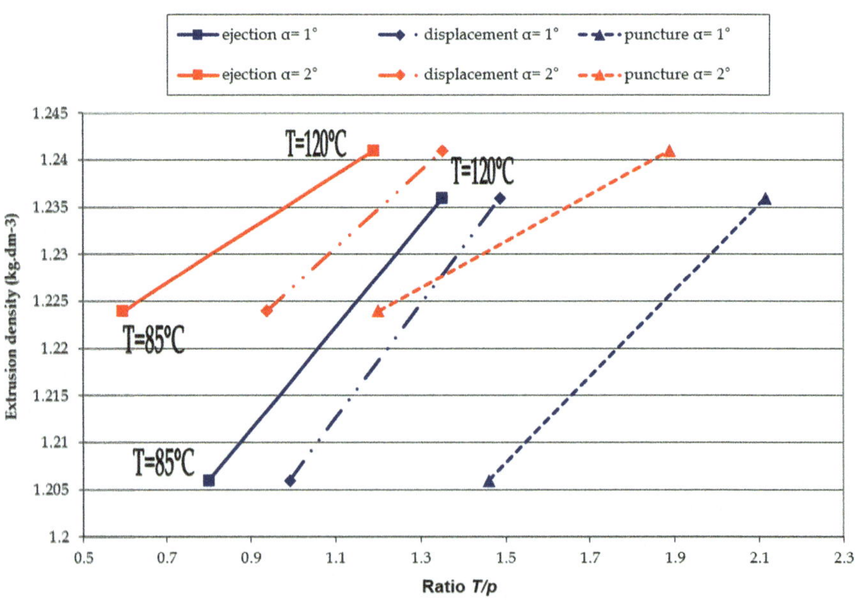

Fig. 6.54 Dependence of extrusion density when pressing with a tapered pressing chamber on temperature to pressing pressure ratio

the point of view of operational economy. By increasing the temperature, it is possible to achieve the desired result both from the point of view of extrusion quality and from the point of view of the economy of production.

6.2.4 Distribution and Action of the Friction Forces Within the Pressing Chamber

In this chapter, the primary goal is not to determine the influence of the selected parameter, but to verify in an available way the distribution and action of friction forces in the pressing chamber. The experimental stand is not suitably equipped to monitor the frictional forces directly. However, if we realize that the friction forces are based on the action of radial forces and the state and conditions of friction pairs, it is possible to record certain interesting findings throughout the experiments. These will be useful in the future when designing a new experimental pressing stand and creating a comprehensive experimental research procedure of the compaction process.

The coefficient of friction, which defines the conditions between the friction pairs, can be influenced in the biomass compaction process in two basic ways. First is by the pressing chamber surface roughness, which in practice disappears over time since biomass and wood sawdust are highly abrasive. The second way is to determine the optimal ratio between technological parameters (pressing temperature, compaction pressure) and the parameters of the raw material (material fraction size and moisture content). All these factors ultimately affect the action of pressures within the pressing chamber and thus impact the resulting density of extrusions. It would be important and useful if a comprehensive large-scale experiment was carried out in the future to determine the effects of the coefficient of friction. However, this involves designing a new experimental pressing stand, which will accommodate a higher variability in design parameters and be equipped with various sensors for monitoring the behavior of the applied pressures over the entire length of the pressing chamber.

Therefore, the aim was to carry out the experiments within the limits of the existing apparatus. Several findings concerning the recorded pressure profiles within the pressing chamber were given in the previous sections and are therefore no longer described. In addition to the abovementioned factors, which influence the action and distribution of pressures within the pressing chamber, the length of the compressed column within the pressing chamber is also a very important parameter. As the length of the compressed column increases, the radial pressures decrease and, conversely, the back pressure increase. This condition is mainly utilized in "open" pressing methods. During the experiments, some knowledge was also obtained regarding the effect of the column length on the friction within the pressing chamber. The following Figs. 6.55, 6.56, and 6.57 show the behavior of pressures required for extrusion as a function of the compressed column length.

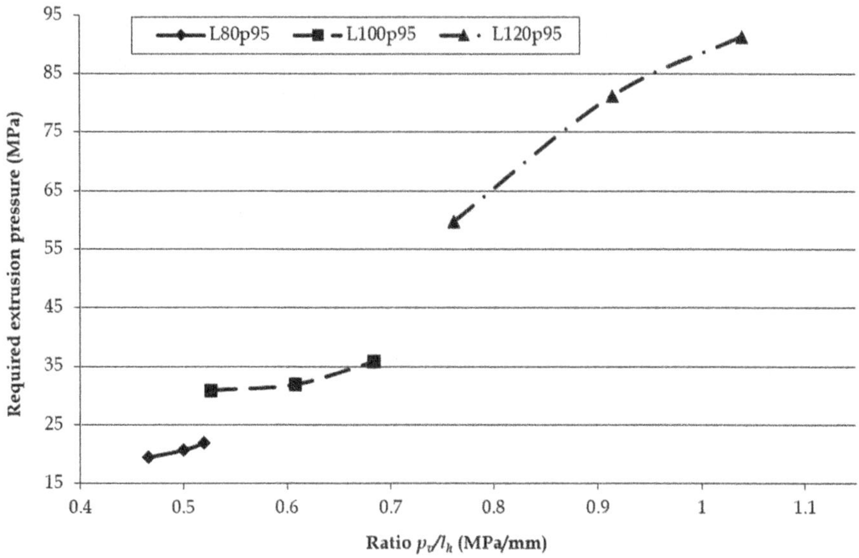

Fig. 6.55 Dependence of the pressure necessary for extrusion on the pressure to column length ratio for varying chamber lengths at 95 MPa of pressing pressure

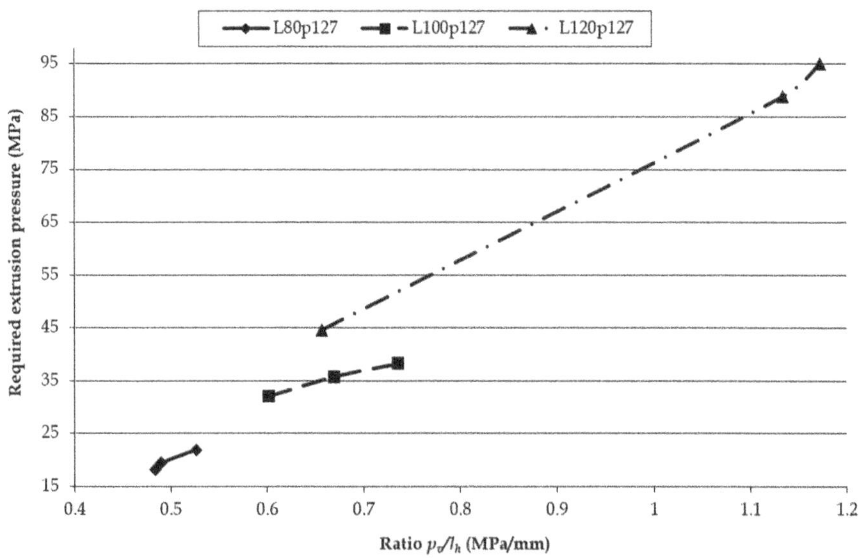

Fig. 6.56 Dependence of the required extrusion pressure on the pressure to compressed column length ratio for different chamber lengths at 127 MPa of pressing pressure

The measured pressures represent the value at which the frictional forces between the pressed material and chamber walls, caused by radial pressure components during extrusion, must overcome. After overcoming these pressures, the pressure required for extrusion decreases. It is assumed that the pressure required for

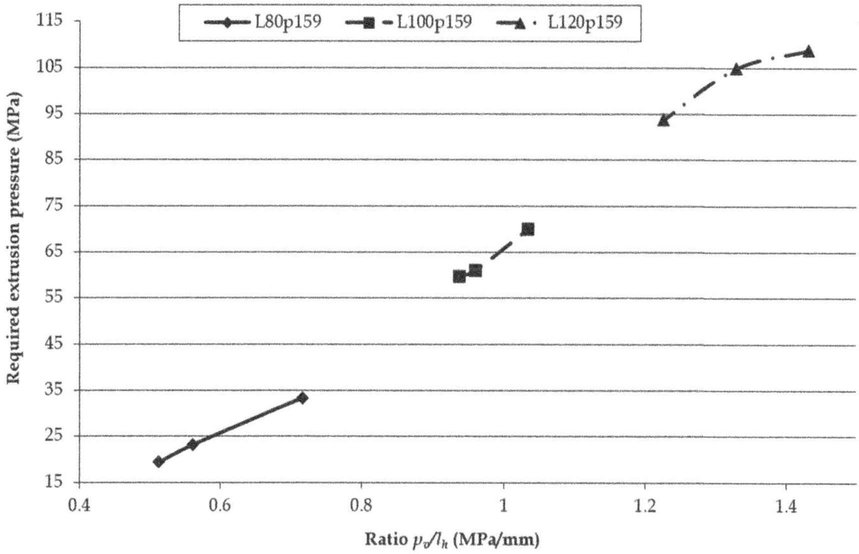

Fig. 6.57 Dependence of the required extrusion pressure on the pressure to compressed column length ratio for different chamber lengths at 159 MPa of pressing pressure

extrusion will depend only on the size of the contact surfaces. However, this pressure also depends on the radial pressures and therefore the necessary pressure required to displace the extrusion out of the pressing chamber is not directly proportional to the column length alone. Based on the results shown within the previous figures, it can be stated that the friction conditions vary throughout the compaction process depending on the pressing pressure, length of the pressed column, and the length of the pressing chamber. Different friction conditions can also be observed depending on the fraction size of the raw material [27]. In general, the dependency of extrusion density on fraction size is known. In Fig. 6.58, it can be seen that as the fraction size of the pressed material decreases and the compaction pressure increases, the density of the extrusions increases. This graphical dependence was created based on the results of an experiment where sawdust at 8% moisture content was compressed. When the fraction size affects the force required for extrusion, the situation is opposite. As the fraction size of the raw material decreases, so does the force required for extrusion [27, 28]. This is due to different friction conditions depending on the fraction size of the raw material.

For clarity, the required extrusion forces are expressed over a 1-mm pressed column (Fig. 6.59). These dependencies clearly demonstrate the effects of fraction size on the distribution and action of frictional forces within the pressing chamber. The interrelationship of the effects of fraction size on the density of extrusions and the force required to displace them within the pressing chamber is shown in the following figure (Fig. 6.60). It should be noted here that in this experiment the compaction pressure value was 15 MPa. With these dependencies, it would be possible, after considering the effects of the remaining parameters, to find the optimal value of the fraction size in terms of the force required for extrusion, for different settings.

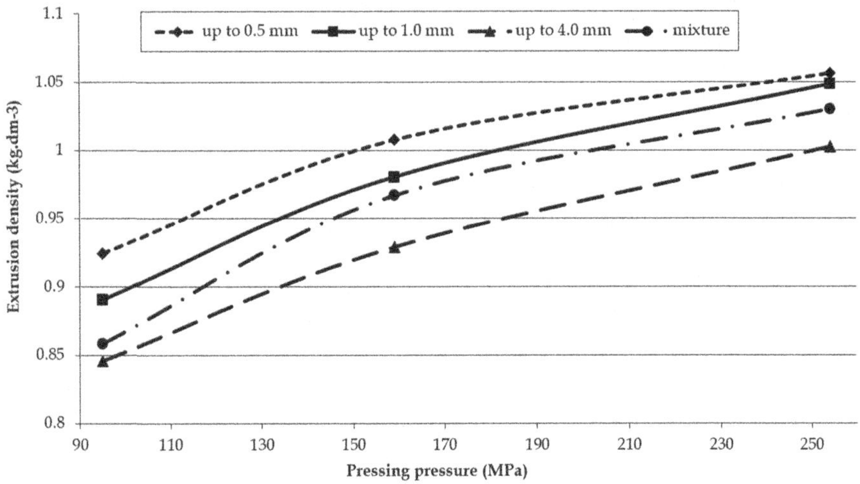

Fig. 6.58 Dependence of extrusion density on the pressing pressure for different fraction sizes of the input material

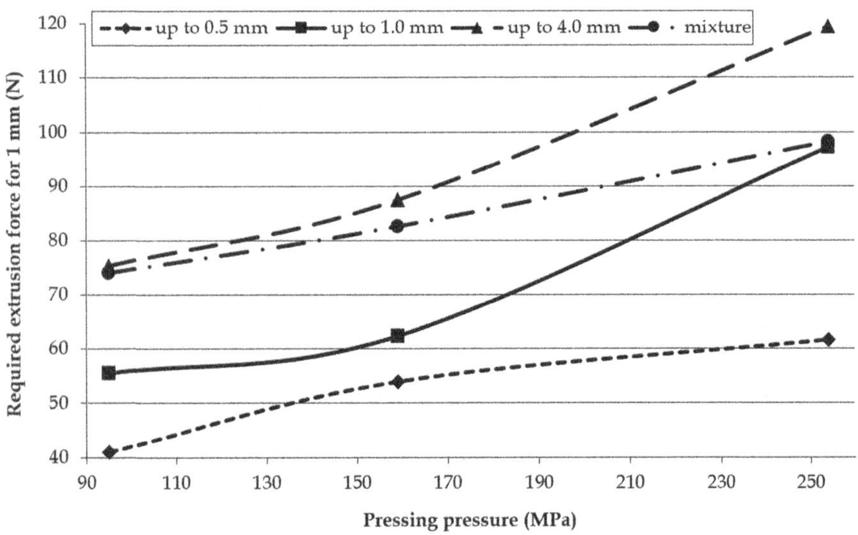

Fig. 6.59 Dependence of required extrusion force on the pressing pressure for various levels of fraction size

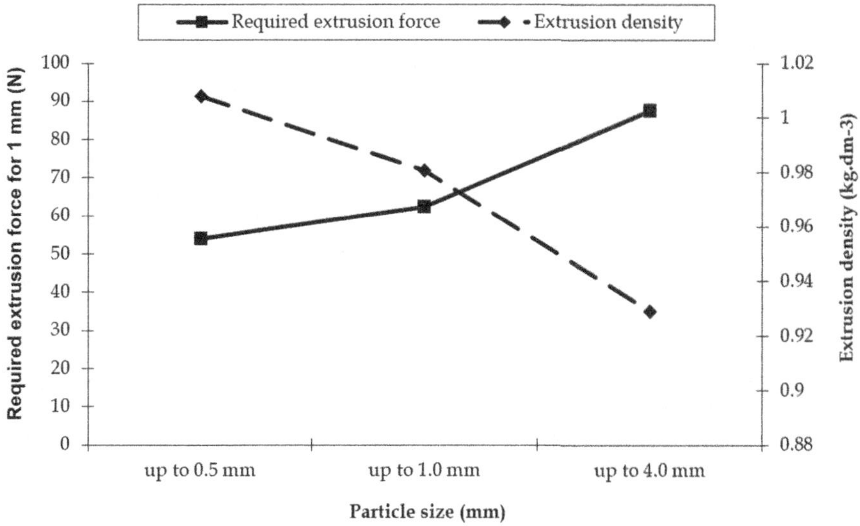

Fig. 6.60 Interrelationship between the particle size and extrusion density as well as the required extrusion force

References

1. Horníková, A.: Design and Evaluation of Experiments with Applications, Bratislava (2009) (in Slovak)
2. Križan, P., Svátek, M.: Research of the process of compaction of different types of biomass, 1st ed. 182 p. FX s.r.o., Bratislava (2014). ISBN 978-80-89313-59-4 (in Slovak)
3. Križan, P., Matúš, M., Svátek, M.: Findings from the research of pelleting of selected species of dendromas and phytomasses. In: Briquetting and Pelleting 2012: A Collection of Lectures from the 6th International Conference. Bratislava, 25.-26.1. 2012, pp. 104–111. STU in Bratislava, Bratislava (2012). ISBN 978-80-227-3641-1 (in Slovak)
4. Križan, P., Svátek, M.: Design of an experiment for compaction process research. In: Proceedings of the ERIN International Conference 2008, Bratislava, 23–24.04.2008, p. II-50. ISBN 978-80-227-2849-2, CD-ROM (in Slovak)
5. Križan, P.: The Process of Pressing Wood Waste, 1. vyd, 197 p. Publishing House STU, Bratislava (2014). ISBN 978-80-227-4251-1 (in Slovak)
6. Kováč, P.: Research into the Influence of Design Parameters on the Final Quality of Moldings. Master thesis. ÚSETM. FME STU v Bratislave, SjF-5254-26779, Supervisor: Peter Križan, 83 p (2010) (in Slovak)
7. Li, Y., Liu, H.: High-pressure densification of wood residues to form an upgraded fuel. Biomass Bioenerg. **19**, 177–186 (2000)
8. Liška, M.: Research into the influence of design parameters on the final quality of compacts during compaction. Master thesis. ÚSETM. FME STU in Bratislava, SjF-20736-58456, Supervisor: Peter Križan, 91 p (2013) (in Slovak)
9. Matúš, M., Križan, P.: Influence of structural parameters in compacting process on quality of biomass pressing. Aplimat J. Appl. Math. **3**(3), 87–96 (2010). ISSN 1337-6365
10. Jarošová, E., Žaludová, A.: Designing Experiments, p. 105. Česká společnost pro jakost (1997). ISBN 80-02-01154-6 (in Czech)

11. Perez, E.: Module – Pellets Press Components (Rolls, Dies, Knives, Feeder). The Agricultural University of Norway, online, cited 05.12.2020. http://docplayer.net/21050493-Module-pellet-press-components-rolls-dies-knives-feeder.html
12. STN 015030/a:1983 Stanovenie zrnitosti sitovým rozborom/Test sieving (in Czech)
13. STN EN ISO 18134-1:2015, Tuhé biopalivá. Stanovenie obsahu vlhkosti. Metóda sušením v sušiarni. Časť 1: Celková vlhkosť. Referenčná metóda./Solid biofuels. Determination of moisture content. Oven dry method. Part 1: Total Moisture. Reference Method (in Slovak)
14. Palenčár, R., et al.: Statistical methods in metrological and testing laboratories, 366 p. Publishing Hourse Grafické štúdio Ing, Peter Juriga (2001). ISBN 80-968449-3-8 (in Slovak)
15. Svátek, M., Križan, P.: Proposal for experimental plan and evaluation methodology for research of densification process of different types of mixtures. In: ERIN 2013. Proceedings of Abstracts with Full Papers: 7th International Conference for Young Researchers and Ph.D. Students. Častá-Papiernička, Slovakia, May 15–17, 2013. Publishing House STU, Bratislava (2013). ISBN 978-80-227-3934-4
16. DIN 51731:1996, Testing of Solid Fuels – Compressed Untreated Wood, Requirements and Testing. Deutsches Institut für Normung, Berlin
17. Križan, P., et al.: Stabilization time of extrusions after briquetting. In: ERIN 2011: Education. Research. Innovation. 5th Annual International Travelling Conference for Young Researchers and PhD Students. Tatranská Kotlina, Slovakia, 13th–16th April 2011. Apeiron EU, Prešov, pp. 127–134 (2011). ISBN 978-80-89347-05-6
18. Križan, P., et al.: Stabilization time as an important parameter after densification of solid biofuels. Acta Polytechnica. **54**(1), s. 35–41 (2014). ISSN 1210-2709
19. DIN 52182 Bestimmung der Rohdichte. Deutsches Institut für Normung, Berlin
20. Kačík, F.: Influence of Temperature and Moisture on Changes of Wood Carbohydrates, Scientific Study 4/1997/A, p. 68. Technical University in Zvolen (1997). ISBN 80-228-0608-0 (in Slovak)
21. Solár, R.: Changes in Lignin in the Processes of Hydrothermal Treatment of Wood, Scientific Study 1/1997/A, p. 56. Technical University of Zvolen (1997). ISBN 80-228-0599-8 (in Slovak)
22. Križan, P., Svátek, M., Matúš, M.: Analysis of the significance of technological parameters at briquetting of selected types of hardwood and softwood. In: Aplimat 2011. Proceedings, pp. 395–404. FX s.r.o., Bratislava (2011). ISBN 978-80-89313-51-8
23. Križan, P., et al.: Difference between compacting of softwood and hardwood. In: Scientific Proceedings 2009: Faculty of Mechanical Engineering, STU in Bratislava. Slovak University of Technology in Bratislava, Bratislava, pp. 161–167 (2010). ISBN 978-80-227-3326-7
24. Nielsen, N.P.K., et al.: Importance of temperature, monture content, and species for the conversion process of wood residues into fuel pellets. Wood Fibre Sci. **41**(4), 414–425 (2009). Society of Wood Science and Technology
25. Stelte, W.: Fuel Pellets from Biomass. Processing, Bonding, Raw Materials. PhD. thesis, Danish National Laboratory for Sustainable Energy, Technical University of Denmark, December 2011, p. 47. ISBN 978-87-550-3599-1
26. Holm, J.K., et al.: Fundamentals of Biomass Pellet Production. Annual Research Report (2005). ISBN 87-74-75325-8
27. Ghadernejad, K., Kianmehr, M.H., Arabhosseini, A.: Effect of moisture content and particle size on energy consumption for dairy cattle manure pellets. Agric. Eng. Int. CIGR J. **14**(3), 125–130 (2012). ISSN 1682-1130
28. Serrano, C., et al.: Effect of moisture content, particle size and pine addition on quality parameters of barley straw pellets. Fuel Process. Technol. **92**, 699–706 (2011)

Chapter 7
Summary of Experimental Results and Research with Conclusions

The present scientific publication presents a basic picture of the interaction between design parameters of the pressing chamber of a pressing machine with respect to extrusion quality throughout the pressing process. With increasing demands for energy self-sufficiency, the requirements for efficient production of solid biofuels are also increasing. The importance of this is also evident with the decreasing suitability of biomass resources (wood biomass, agricultural phytomass, etc.). Recently, there has been a trend to broaden the reach of energy extraction, and thus use of lower quality materials with a lower content of cellulose, hemicellulose, and lignin. Unfortunately, most of these materials do not benefit from the optimization of material (moisture and fraction size) and/or technological parameters (pressing pressure and temperature). It is in such cases that optimization of the design parameters of the pressing chamber for a particular type of compaction material must be considered.

At first glance, the compaction of biomass is a very interesting, simple, and clear process. However, if we address the issue in more detail, we will find that the compaction process can be characterized by a large variety of influencing parameters and it is not easy to optimally solve the problems that arise during the compaction of biomass. This monograph indicates possibilities for optimizing the construction of a pressing machine with respect to the energy efficiency of the process and resulting quality of extrusions. Despite the results obtained and presented, however, it is necessary to continue experimental research.

7.1 Practical Use of the Experimental Research Results

Every designer must adhere to certain knowledge when designing a new machine or upgrading existing machines and machine parts. On the one hand, knowledge of particular structural nodes and parts of the machines themselves or the process that

P. Križan, *Biomass Compaction*, https://doi.org/10.1007/978-3-030-89956-1_7

will take place must be obtained. If the acquired knowledge is complete and accurate, it can be very useful and suitably used in optimizing the design of equipment, the resulting quality of production, and improve the performance of the press while minimizing its power requirements. These attributes also significantly affect the economic balance of the production line. Manufacturers of compaction machines, however, do not pay enough attention to the scientific knowledge of the process itself, and the design of pressing mills is based mainly on "experience" by means of trial and error. This prolongs the process of designing and manufacture equipment and ultimately makes the technology and products itself needlessly expensive. The fact that compaction machines, in a way, help to protect the environment by recycling various types of waste prioritizes their use. This is the main reason why a debate has recently opened in this field, with a movement to support research, develop and widen the use of pressing machines and their products.

Our institute strives to contribute to knowledge that will be useful in the optimization of pressing machine design and the process itself. This monograph presents findings and results that, to some extent, clarify the processes and events taking place within the pressing chamber of a pressing mill in the compaction of biomass. In order to be able to make recommendations for the design and construction of pressing machines, results based on many measurements on relevant equipment must be available. The findings presented in this publication are mainly based on results obtained from experiments performed on an experimental pressing stand. The experiments were aimed at investigating and determining the influence of pressing chamber design parameters on the final quality of extrusions.

The specific benefits of this monograph can be summarized in the following points:

1. Theoretical analysis of the construction parameters of a pressing machine and their influence on the quality of extrusions. Specification of design parameters that affect the resulting quality of extrusions (density) for the compaction of biomass.
2. Theoretical analysis of pressing chamber design parameters and their influence on extrusion quality based on a mathematical analysis and quantification of the significance of monitored parameters. Specification of the design parameters of the pressing chamber, which affect the final quality of the extrusions.
3. Establishment of an experimental procedure with experimental research performed to determine the influence of the monitored pressing chamber design parameters on the final quality of extrusions. The proposed experimental procedure may also be used for future experimental research.
4. Investigation of the pressing method, chamber length, and conicity and its influence on the final quality of extrusions.
5. Definition and quantification of the significance of friction between the input material and the walls of the pressing chamber during the compaction of pine sawdust.
6. Identify and define the significance of the experiments performed with respect to the optimization of pressing chamber construction of pressing mills.

In general, it can be stated that the set goals were met, while this publication is a summary of some of the knowledge gained from laboratory measurements and theoretical analysis performed at ISETM FME STU in Bratislava. The issue of biomass compaction today represents a broad field of knowledge with a large number of topics for experimental research, in the form of basic and applied research.

Ultimately, it can be stated that the knowledge gained from the presented experimental research on the pressing chamber design parameters was also used in solving projects under the Operational Program for Research and Development for the development of progressive biomass compaction technology and production of highly productive machines, supported by the European Structural Fund. One of the goals of this project was to design briquetting press using a screw as a pressing tool (Figs. 7.1 and 7.2). The findings from this experimental research were used in the design of the press chamber and in determining the ranges of operating parameters for the machine. After manufacture and assembly of the machine, the initial experiments were successfully carried out resulting in high-quality briquettes (Fig. 7.3).

The primary benefit of the experimental research was the acquired knowledge, which is valuable mainly in the pedagogical, design, and construction process. From this point of view, it is clearly important to continuously engage in scientific research activities in this field. However, the knowledge gained from the experimental research is used at our workplace in solving smaller problems, during various collaborations with companies throughout the industry. To document the practical use of the acquired knowledge, it is possible to give some examples of cooperation with companies from practice, in which the knowledge gained by experimental research of the pressing chamber design parameters was directly utilized:

A Czech company producing sunflower oil approached us with cooperation in compacting their waste—sunflower husks. The intention was to produce pellets from sunflower husks, which would then be either energy-enhanced or enhanced in

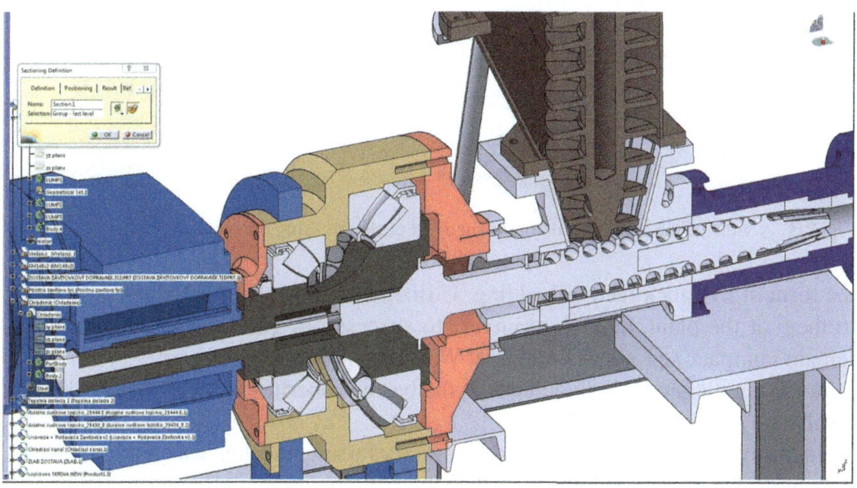

Fig. 7.1 Section of the screw press [1, 2]

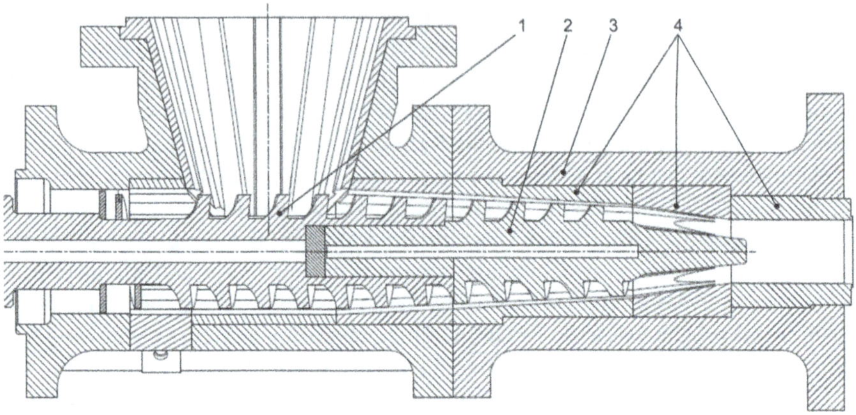

Fig. 7.2 Parts of the screw press (1—feed screw, 2—pressing screw, 3—pressing chamber, 4—nozzle) [1, 3]

Fig. 7.3 A view of the extrusion being pushed out of the pressing chamber [4]

the form of livestock feed or bedding. Utilizing their standard pellet press available to them at the plant, they were unable to successfully produce pellets that would remain compact and not disintegrate. We subjected supplied samples of sunflower husks to simple analyses to determine the basic properties of sunflower husks. The results are shown in Table 7.1.

In the experimental research and laboratory testing, the goal was set to define and determine the degree of influence that the technological and material parameters had on the final quality of extrusions and their dilatation. If it was not possible to increase the mechanical indicators of extrusion quality by adjusting the aforementioned parameters, the effects of the design parameters would be tested in the laboratory in order to optimize the shape of the pressing chamber.

Prior to experimental research, two samples of material with different fractional compositions were prepared. The first sample of material was in the "raw" form (untreated fraction composition—coarse fraction). The second sample was subjected to fraction refinement. To adjust the fractional composition of sunflower husks, the Stoza ŠV 5 hammer mill utilizing a sieve with mesh size ϕ 4 mm was suitably selected and used. It is also suitable to use a knife mill to adjust the size of the sunflower husks. For the production of pellets with a diameter of 6 mm, a larger diameter of sieve openings was not recommended. With a larger fraction of crushed husks, the mechanical quality indicators of the pellets decreased. The size of the supplied sunflower husk fraction was suitable for the pelleting experiments in the form of pellets with a diameter of 6 mm. Crushing was performed in order to determine the influence of the fraction composition and the influence of the fraction size on the final quality of the extrusions. Sample before crushing—the coarse fraction (supplied sample)—and after crushing (fine fraction) can be seen in Figs. 7.4 and 7.5. Subsequently, the fractional composition was determined by particle size analysis on a Retsch vibrating screen. The results are shown in Fig. 7.6. It was obvious

Table 7.1 Properties of the sunflower samples [5]

Observed parameter	Gross fraction
Relative moisture (%)	9.24
Bulk density (kg/dm^3)	0.1614

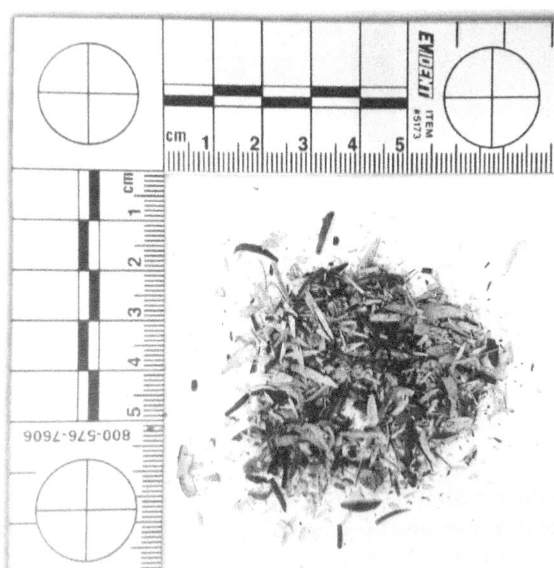

Fig. 7.4 Sunflower husks—sample (gross fraction) [5]

Fig. 7.5 Sunflower husks—shredded fraction (fine fraction) [5]

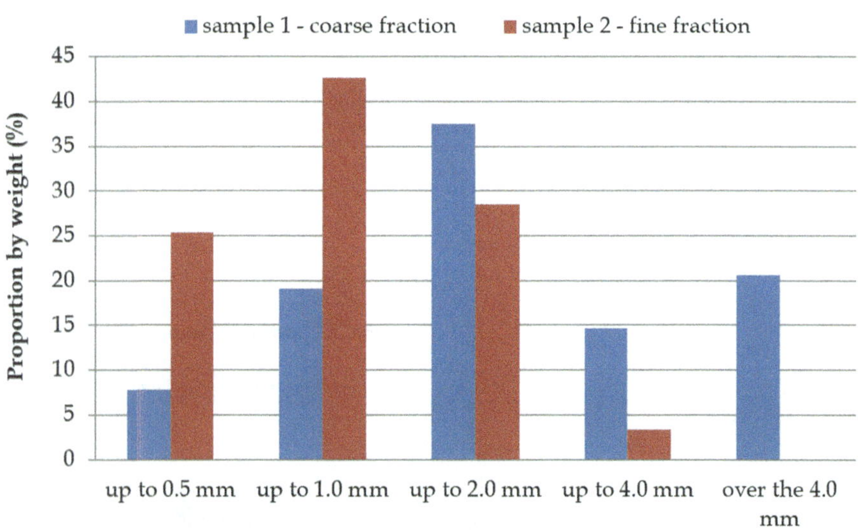

Fig. 7.6 Granulometric analysis and proportion of individual fractions throughout the volume [5]

that the crushing of the fraction caused a change in the proportions of individual fractions throughout the volume, whereby the volume of finer fractions increased. From the point of view of extrusion quality and compaction technology, fractions with a proportion of finer fractions are more suitable and can significantly contribute to increasing the quality of extrusions.

Samples prepared in this way were subject to experimental research on a pressing stand, which was available at our workplace. In order to determine the effect of the change in pressing pressure, an experiment was performed where the pressing pressure would be increased within the range of 63 MPa to 191 MPa. In order to determine the effect of the change in pressing temperature, an experiment was carried out in laboratory testing without the use of additional heating (i.e., at ambient temperature of 23 °C) followed by inducing heat of 100 °C. From a technological point of view, it is not important to carry out the experiment outside the stated intervals. The extrusions were weighed and measured for basic dimensions (diameter and length) immediately after extrusion from the pressing chamber, while the specific weight of the extrusions was calculated, which is one of the basic indicators of the quality. After the extrusions stabilized, i.e., duration where the dilatation of the extrusions had stabilized, they were reweighed and measured. In this way, it is possible to generally define the expansion of the extrusions and determine the influence of the mentioned technological parameters on the expansion of the extrusions. The processed results are shown in the following graphical dependencies.

The pressing chamber of the experimental stand has a pressing channel length of 140 mm and diameter of ø 20 mm. Therefore, it is possible to produce extrusions with a diameter of ø 20 mm. Extrusions made throughout laboratory testing are shown in Figs. 7.7 and 7.8.

From the results (Fig. 7.9), it is clear that with increasing compaction pressure, the resulting density of the compacts also increases. However, it can also be seen here that the dilatation negatively affected the size of the extrusions, which was reflected in a decrease in density. On average, the density of extrusions decreased by 4.82% after stabilization. When pressing with additional heating (100 °C), it was

Fig. 7.7 Extrusions from sunflower husks—pressing at room temperature [5]

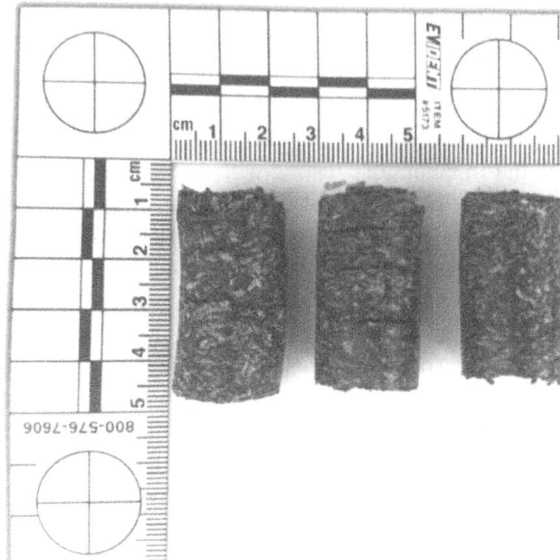

Fig. 7.8 Extrusions from sunflower husks—pressing at 100 °C [5]

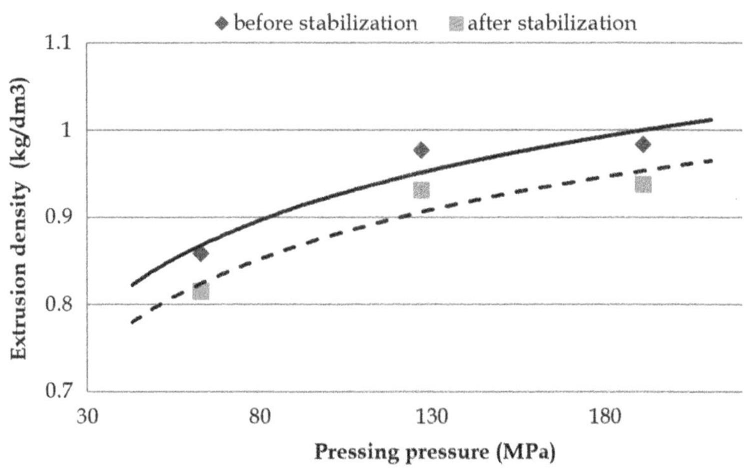

Fig. 7.9 Dependence of extrusion density on the pressing pressure without additional heating [5]

also noted that as the compaction pressure increased, the resulting density of the extrusion also increased (Fig. 7.10). However, the additional heat had a positive effect on the extrusions during the stabilization period and mitigated the effect of dilatation. After stabilization, higher values of extrusion densities were measured, on average, by 3.39%. The additional heat had a positive effect on the expansion of the extrusions.

The results shown in Fig. 7.11, however, show that when comparing the densities of extrusions obtained by pressing with and without additional heating, the values

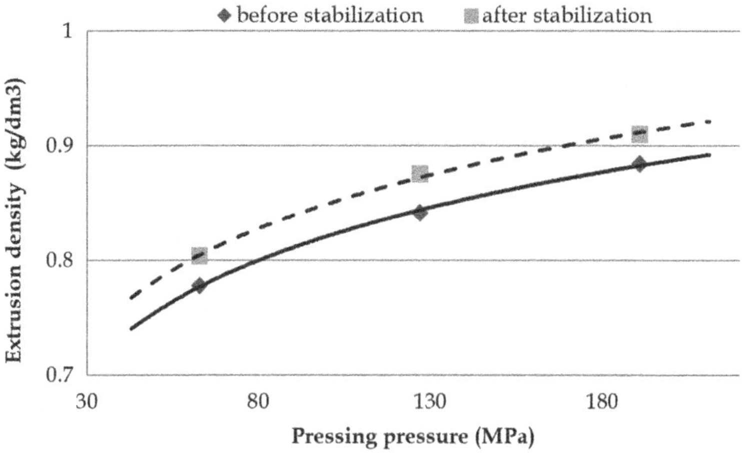

Fig. 7.10 Dependence of extrusion density on the pressing pressure with additional heating at 100 °C [5]

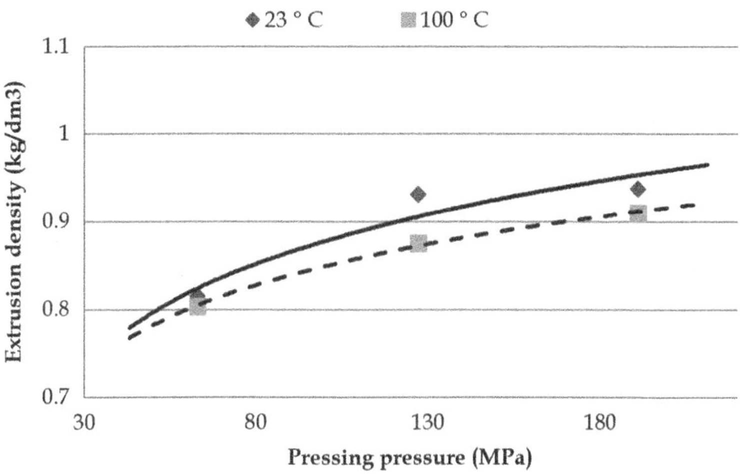

Fig. 7.11 Dependence of extrusion density on the pressing pressure at different heating values [5]

obtained within the pressing without additional heating were better. This fact can be attributed to the chemical composition of sunflower husks and the nature of this material. With low contents of lignin and hemicellulose in the compressed material, the additional temperature during compaction has no significant effect on the resulting density of the extrusion. It should be noted that the dependences shown in Fig. 7.11 are after stabilization of the extrusions. On average, the values of the extrusion densities obtained by pressing without additional heating were higher by 3.57% compared to the values obtained in compaction with additional heating of 100 °C.

The extrusions showed compactness, the binding of sunflower husk particles in the cross section of the extrusion (Fig. 7.12) appeared to be sufficient. No significant effect was observable when changing the size of the fraction and pressing temperature. Since only the change in the compaction pressure had a more significant effect, it was necessary to test the pelleting of sunflower husks on a pelletizing press. However, on standardly produced pellet mills, there is only a limited possibility of compaction pressure, i.e., the interval of change (in this case increase) in the compaction pressure. According to the client, they carried out unsuccessful pelleting at maximum pressures using a die with cylindrical holes. Thus, the only possibility was to change the shape of the die holes in order to change the pressure distribution, which should result in better compaction. In our own pelleting experiments, we used a pelletizing die with conical holes. Due to the protection of intellectual property, it is no longer possible to specify the shape and size of the press holes of the pellet die. Subsequently, the produced pellets (Fig. 7.13) were subject to laboratory testing and determination of mechanical quality indicators and some chemical–thermal quality indicators. The results are summarized in the following Table 7.2.

Based on the above results, it can be stated that sunflower husks can be pelletized to achieve a satisfactory quality of extrusion, using a pellet die with conical holes. The refinement of the fraction and the change in moisture have proven to be negligible, and therefore not necessary to carry out such mechanical treatments of the bulk material before the pelleting process for sunflower husks. The expected effect in the form of increased pellet quality from fine fractions was not confirmed and seemed to indicate the contrary.

A similar assignment from practice was addressed to our workplace by a Slovak company producing chocolate. Based on the assignment, the goal was to effectively mechanically process the company's technological waste—cocoa shells into a form more efficient for long-distance transport and possibly use as bedding for livestock. The aim of the expert analysis was to verify the possibility of using compaction technology in the processing of cocoa husks, to define the technological parameters of compaction technology, which influence the final quality of extrusions and to determine the influence of mechanical treatment of the input material (cocoa husks)

Fig. 7.12 Compacted particles of the extrusion (sunflower husks) [5]

Fig. 7.13 Pellet made from sunflower husks [5]

Table 7.2 Properties of pellets made from sunflower husks [5]

Observed parameter	Client pellet	FME pellet
Moisture content (%)	8.61	5.60
Combustion heat (MJ/kg)	–	19.07
Bulk density (kg/dm^3)	0.3216	0.61054
Particle density (kg/dm^3)	0.6558	1.237
Mechanical durability (%)	61.63	87.08
Abrasion (%)	32.79	12.63
Pellet hardness (N)	137.5	207

on the final quality. Supplied (non-crushed) samples of cocoa husks were subject to simple analyses to determine the basic properties. The results are shown in Table 7.3.

The procedure for mechanical treatment and sample preparation was similar. Again, the sample was refined before laboratory testing and trials. Two samples of material with different fractional compositions were prepared by crushing. The first sample of material to be tested and pelletized was in the untreated fraction "as supplied" by the client (coarse fraction). The second sample was refined by crushing (fine fraction). The Stoza ŠV 5 hammer mill with a ø 4-mm sieve was suitably selected and used to adjust the fractional composition of the cocoa shells. The size of the supplied cocoa shell fraction was suitable for the required pelleting experiments in the form of pellets with a diameter of 6 mm, but only due to the nature of this material. Crushing was performed in order to determine the influence of the fraction composition and the influence of the fraction size on the final quality of the extrusions. Sample before crushing—the coarse fraction (supplied sample)—and after crushing (fine fraction) can be seen in Figs. 7.14 and 7.15. Subsequently, the

Table 7.3 Properties of the supplied material samples [6]

Observed parameter	Coarse fraction (supplied sample)	Fine fraction (crushed samples)
Moisture content (%)	5.723	5.428
Bulk density (kg/dm^3)	0.15055	0.44363
Combustion heat (MJ/kg)	16.5912	–

Fig. 7.14 Cocoa shells—supplied (coarse fraction) [6]

Fig. 7.15 Cocoa shells—crushed (fine fraction) [6]

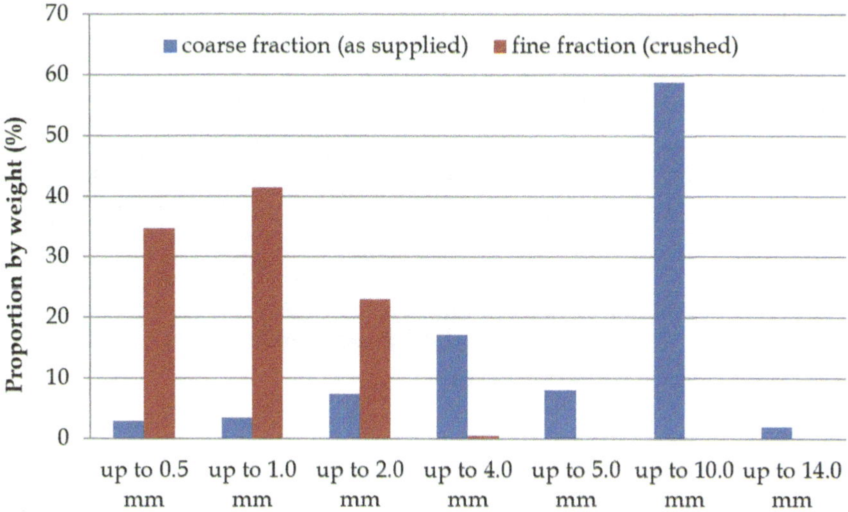

Fig. 7.16 Granulometric analysis and proportion of individual fractions throughout the volume [6]

fractional composition was determined by particle size analysis using a Retsch AS 200 vibrating screen. The results are shown in Fig. 7.16. It is obvious that crushing the shells caused a change in the proportions of individual fractions throughout the volume where the proportion of finer fractions increased. In terms of the quality of extrusions and compaction technology, fractions with a larger proportion of finer fractions are theoretically more suitable and can significantly contribute to increasing the quality of extrusions.

As can be seen from the above figures and also from the results of the particle size analysis, crushing the input cocoa shells caused a significant change in the proportions of individual fractions throughout the volume. This was also reflected in the change in parameters of the material sample (see. Table 7.3). The relative humidity of the sample also decreased slightly when crushed, but this was not a significant difference. A more significant difference was observed when determining the bulk density. The bulk density of the fine (crushed) sample increased 2.94-fold compared to the coarse (as delivered) sample. So it can be stated that by crushing the material beforehand can significantly reduce the transport costs. An illustration of this difference can be observed in Figs. 7.17 and 7.18. Significant differences in the volume of 200-gram samples for both fraction sizes can also be observed.

The next phase was laboratory testing of cocoa shells in terms of their "pressability." This phase is a necessary before the actual compaction tests, because the basic goal of said testing is to confirm or reject the suitability of the compaction technology for processing cocoa shells. It is also possible to determine the technological parameters of the compaction process and their influence on the final quality of extrusions and their dilation by means of laboratory testing. Laboratory testing was performed on an experimental press stand. The aim of the laboratory testing

Fig. 7.17 200 g of coarse fraction

Fig. 7.18 200 g of fine fraction

was therefore to verify the suitability of compaction technology for the processing of cocoa shells and to determine the impact of changes in technological parameters on the resulting quality of extrusions and their dilatation. Extrusions of cocoa shells from both fractions can be seen in Figs. 7.19 and 7.20.

In order to determine the effect that the change in compaction pressure has, a pressing experiment was performed where the pressure was regulated from 63 MPa to 222 MPa. Due to the nature and structure of the input material, it was not important to experimentally determine the effect of the change in pressing temperature. Our experience in pressing materials with a similar chemical composition to the cocoa husks in question, i.e., materials with a low content of lignin and hemicellulose, shows that the additional temperature does not have a significant effect on the resulting extrusions density throughout compaction. Thus, all experimental tests were performed without the use of additional heating (i.e., at ambient temperature of 23 °C). From a technological standpoint, it is not important to carry out an experiment outside the stated intervals. The extrusions were weighed and measured for basic dimensions (diameter and length) immediately after extrusion from the chamber. The results are shown in the following graphical dependencies.

Density was calculated from the measured values of the extrusions diameter, length, and weight. Subsequently, graphical dependencies were constructed, which

Fig. 7.19 Extrusions of cocoa shells—coarse fraction

Fig. 7.20 Extrusions of cocoa shells—fine fraction

are indicators of the results. From the above results (Fig. 7.21), it is clear that as the compaction pressure increases, the resulting density of the extrusions also increases. However, it can also be seen here that the dilatation negatively affected the size of the extrusions, which resulted in a decrease in density. On average, the density of extrusions decreased by 14.53% after stabilization. When pressing material with coarse fractions (uncrushed) (Fig. 7.22), it was again confirmed that the resulting extrusion density increases with increasing compaction pressure. However, it was again observed that the dilation negatively affected the size of the extrusions, reflected in a decrease in their density. On average, the density of the compacts after stabilization decreased by 7.66%, which is a lower dilation effect than with extrusions made with input material of finer fraction.

The change in density after the stabilization time is clearly shown in Fig. 7.23. A negative change is observed for both pressings of fraction sizes of the cocoa husks. However, the coarse fraction (uncrushed) cocoa shells displayed less of a dilation effect, which affected the density due to the increase in compaction pressure. Observations of the results of pressing with fine fraction (crushed cocoa shells) do not display such a trend. The difference is due to the different structures and different representations of individual components of the fractions within the pressed volume of material. This fact is important information when designing the pressing machine and dimensioning the pressing tool. The results shown in Fig. 7.24 show

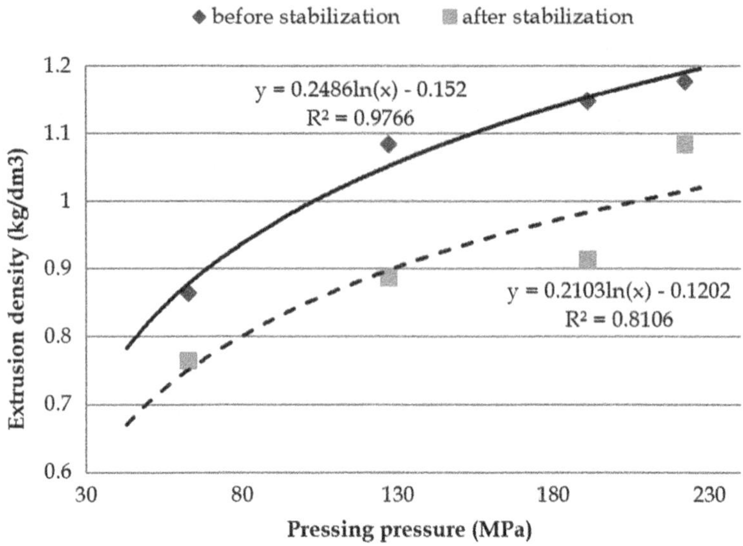

Fig. 7.21 Dependence of the extrusion density on the pressing pressure of input material with fine (crushed) fractions [6]

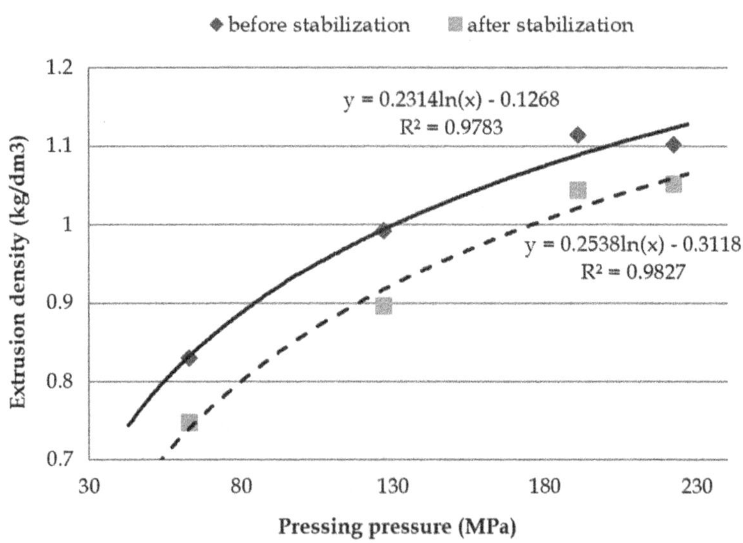

Fig. 7.22 Dependence of extrusion density on the pressing pressure when utilizing input material with coarse (as supplied) fractions [6]

that when comparing the extrusion densities obtained by compacting fine and coarse fractions, the values obtained in the compaction of the coarse fraction were better. This fact can be attributed to the formation of stronger bonds between material particles within the coarse fractions. It should be noted that the dependences shown in

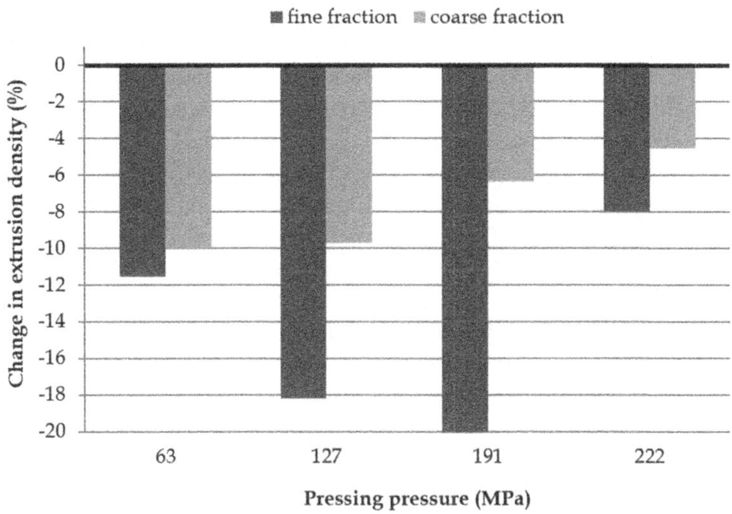

Fig. 7.23 Dependence of the change in extrusion density on the pressing pressure for both fraction sizes

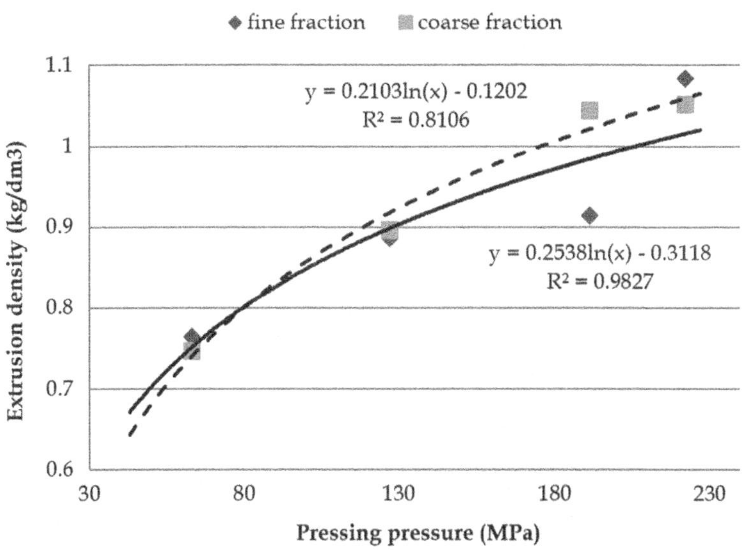

Fig. 7.24 Dependence of the extrusion density on the pressing pressure for both fraction sizes [6]

Fig. 7.24 are after the stabilization of the extrusions, i.e., in a state which can be evaluated "in the field." On average, the values of the extrusion densities obtained by pressing with the coarse fractions were higher by 1.99% compared to the values obtained by pressing with fine fractions.

From these results, it was confirmed that cocoa shells were suitable for compaction technology. It was also interesting to investigate the influence of the compaction pressure and find that refinement (crushing) of the input material was not a necessary part of the design. The results from the tests showed that higher extrusion densities were obtained when utilizing the coarse (uncrushed) fraction.

On the basis of the obtained results, the clients planned to purchase and subsequently operate the pressing technology within a production line, but there remained the question whether to use briquetting or pelleting methods. In order to determine the suitable design of technology for the compaction of cocoa shells, operational tests of cocoa shell briquettes were carried out first. These tests were performed on a hydraulic briquetting press BrikStar 200 and also on a mechanical crank press BL 50-350. The differences between these types of pressing machines have been sufficiently described within Chaps. 2 and 3.

The difference in the type of pressing, i.e., pressing in a "closed" chamber (hydraulic press) compared to pressing in an "open" chamber (mechanical press), was clearly seen. Despite the various possibilities of the press and our efforts, the operational tests on the hydraulic press were not successful. The briquettes did not fully compress and disintegrated immediately after leaving the pressing chamber. The cause can be found in the properties of the briquetted material and the technological parameters, which with regard to the size of the produced briquette (ø 55 mm) the optimal values could not be achieved. Figure 7.25 shows the remains of briquettes after pressing the cocoa husks. For this reason, these extrusions could not be tested or verified.

A few extrusions were attempted on the mechanical crank press and subsequently subjected to the determination of specific weight—density of briquettes (Fig. 7.26). The average achieved value of specific weight was 0.8516 kg/dm^3. Although the test

Fig. 7.25 Failed extrusions of cocoa shells utilizing a hydraulic briquetting press

Fig. 7.26 Briquettes made from cocoa shells produced on a mechanical briquetting press [6]

was "successful" when utilizing the mechanical briquetting press, it can be said that the resulting briquetting in generally was not sufficient to thoroughly compact the cocoa shells. Based on the achieved results utilizing briquetting tests, it is important to state that it is necessary to deal with a thorough optimization of the design parameters of the briquetting chamber depending on the technological parameters of the briquetting press. Under the given conditions, this is the only way to improve the pressure conditions within the pressing chamber.

The cocoa shells were then tested in pelleting methods. These operational tests were performed on a KAHL 33-390 pellet press. The aim of the pellet tests was to verify the suitability of using this technology under standard operating conditions (technology parameters and tool shape) to produce pellets with the best possible output quality. A pelletizing press with a die geometry intended primarily for the pelleting of wood biomass was used. The pellet diameter was ø 6 mm. Palletization tests utilizing the cocoa shells of both coarse and fine fractions were carried out without any major technical or technological complications. In fact, the process went very smoothly. However, the achieved physical–mechanical indicators of pellet quality were at an average to lower level, which can be attributed to the properties of the pelletized material itself—mainly its chemical composition (lignin content, hemicelluloses, etc.). Figures 7.27 and 7.28 show the produced pellets.

The pellets produced were subjected to laboratory testing to determine the mechanical quality indicators and some chemical–thermal quality indicators. The results are summarized in Table 7.4.

Based on the results, it can be stated that cocoa shells may in fact be pelleted with a satisfactory quality of production utilizing pelleting methods. The refining (crushing) of the fraction proved to be unnecessary. The expected effect that increased pellet quality would be achieved from finer fractions was not confirmed. The basic mechanical indicators of pellet quality (bulk density, specific gravity, mechanical resistance, abrasion, and hardness) were determined at approximately the same values for both fraction sizes. The bulk density of pellets produced with coarse

Fig. 7.27 Cocoa shell pellets—coarse fraction [6]

Fig. 7.28 Cocoa shell pellets—fine fraction [5]

fractions was approximately 1.43 times higher than the bulk density of those from crushed fractions. The bulk density of the supplied sample of cocoa shells can be increased by 4.23-fold by pelleting. The pelleting technology achieved higher quality extrusions than the briquetting technology due to the type of construction principle employed as well as the influence of the shape and size of the pressing holes.

Table 7.4 Properties of the pellets produced from cocoa shells [5]

Observed parameter	Coarse fraction	Fine fraction
Moisture content (%)	7.609	7.276
Combustion heat (MJ/kg)	18.5856	18.6141
Bulk density (kg/dm^3)	0.63654	0.63601
Particle density (kg/dm^3)	1.251	1.238
Mechanical durability (%)	68.99	70.58
Abrasion (%)	38.32	46.15
Pellet hardness (N)	158	162.5

Fig. 7.29 Prototype of a wastewater pressing machine (left), briquettes made from wastewater products (right) [7, 8, 10]

The pellet die appeared to be the better solution in terms of the conditions for pressing resulting in better bonding between material particles of the cocoa shells, thus achieving higher quality extrusions.

Even in this case, it has been confirmed that knowledge in the design of pressing machines, in particular the effects of the design parameters of the machine, is useful in cases where the change in material and technological parameters is not sufficient. Knowledge and information regarding the influence of the design parameters were employed at our workplace in other cases as well. Compaction of sludge from a wastewater treatment plant, based on an assignment from a Czech company, could only be realized after structural modification of the pressing chamber and pressing tool [7–10]. Figure 7.29 shows a developed prototype device for sludge compaction from a wastewater treatment plant.

Similarly, by modifying the pressing chamber, satisfactory results could be achieved when briquetting municipal waste on the basis of an assignment from a Slovak company [11–13]. The design of briquetting machines for processing municipal waste can be seen in Fig. 7.30.

For the pelleting of waste from a biogas plant, it was possible to produce pellets (Fig. 7.31) of satisfactory quality by means of a pelleting die with conical holes [14,

Fig. 7.30 Modified briquette press (left), briquette from municipal waste (right) [11–13]

Fig. 7.31 Biogas waste product (left), pellets made from biogas waste (right) [14, 15]

15]. The pellets achieved about 32% better mechanical properties this way when compared to a die with regular cylindrical holes.

All of the mentioned cases are proof of the practical use and benefits of the experimental research and their results. For this reason, I clearly believe that the issues addressed herein display great potential in addressing the efficiency of the compaction process. From this perspective, it is clearly important to continuously engage in scientific research activities within this field. The implementation of comprehensive experimental research should provide useful knowledge regarding the design parameters of pressing machines and their influence on the final quality of extrusions, which should significantly affect the process of optimizing pressing machine design, the pressing tools, and compaction process itself.

7.2 Further Research

Based on the knowledge and experiences gained by the experimental research in this field, it is possible to suggest areas worthy of further research. Despite the experimental research performed on the pressing stand available to us, there is still room for research to determine more complex interactions within the pressing process. Based on the desired parameters, it will be necessary to perform research activities dealing with the following issues:

1. *Design and manufacture a new experimental pressing stand* which will allow for a greater degree of variability in the monitored parameters. The experimental pressing stand currently available at our institute was not specifically designed to monitor construction parameters. The results expressed within this monograph show the need for a more sophisticated apparatus. The new pressing stand should allow for a higher degree of variability, which would allow for simple changes to be made while obtaining more precise results. One possibility would be to use pressing "rings" arranged for a particular chamber configuration, which would allow for quick changes to the pressing chamber (see Fig. 7.32). These pressing rings can be easily mounted and exchange depending on which part of the chamber is necessary to monitor. Their configuration can allow for cylindrical arrangements and conical arrangements. Their use could give valuable information with regard to the pressing chamber geometry and behavior of pressures within the chamber based on the configured geometry. The chamber could be modified for variable changes in nozzle shape. Each ring can be equipped with a transducer, which would provide feedback regarding the pressures occurring throughout the chamber (Fig. 7.33) with respect to the parameters of the press.
2. *Describe the behavior of pressures within the conical pressing chamber precisely and in detail.* Experimental research on the conicity of the pressing chamber (tapper of the chamber walls) should consider a more complex (precise) mathematical model. The model should describe the behavior of pressures within the conical chamber based on changing parameters (chamber length,

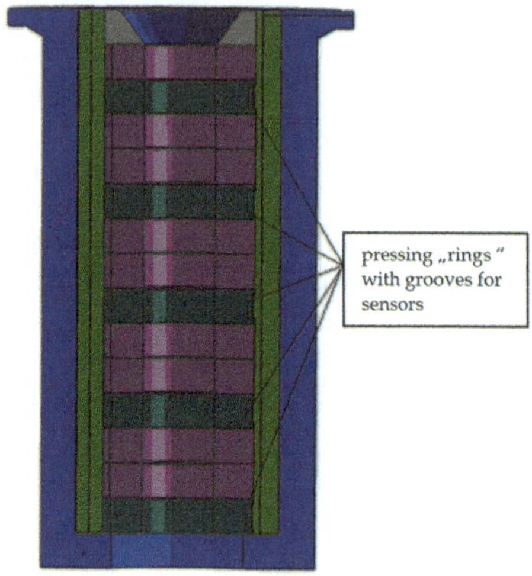

Fig. 7.32 Arrangement of pressing rings forming an arbitrary pressing chamber

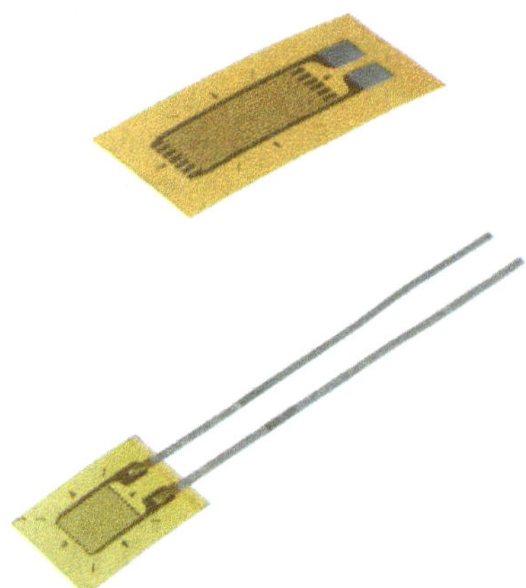

Fig. 7.33 Tensometer wafer

taper length of the conical part of the chamber, taper angle, and coefficient of friction) and evaluate them based on the quality indicators of extrusions.

Based on the information given within this text, it would be possible to monitor the distribution of pressures at different parts of the pressing chamber (Fig. 7.34). Mounting the sensors in, for example, the nozzle would allow for the behavior of the back pressure to be examined.

1. *Describe the behavior of pressures within the conical pressing chamber precisely and in detail.* Experimental research on the conicity of the pressing chamber (tapper of the chamber walls) should consider a more complex (precise) mathematical model. The model should describe the behavior of pressures within the conical chamber based on changing parameters (chamber length, taper length of the conical part of the chamber, taper angle, and coefficient of friction) and evaluate them based on the quality indicators of extrusions.

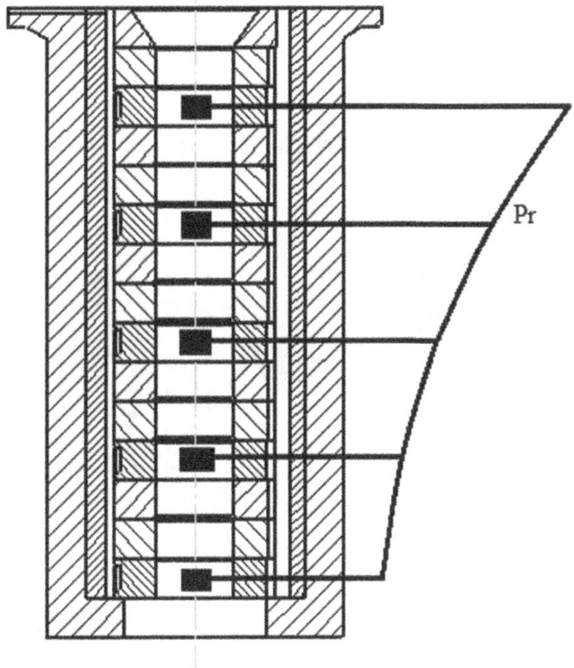

Fig. 7.34 Theoretical distribution of radial pressures throughout the pressing chamber and the sensing points throughout the "ringed" chamber

2. *Perform experimental research to determine the magnitude of the radial pressures* along the entire length of the chamber and *the corresponding magnitude of the back pressure*. The results of this experiment could answer questions regarding the optimization of the chamber size and shape with respect to the quality of extrusions.
3. *Expand the mathematical model*, describing the interdependent interactions of the construction parameters *based on the type of material being pressed*. It is obvious that, apart from the technological and construction parameters, the type of material being pressed affects the distribution and action of pressures within the pressing chamber.
4. *Verify future mathematical models*, describing the effects of interdependent interactions of the construction parameters and real conditions. At our institute, we were able to construct a production line in laboratory conditions; see Figs. 7.35, 7.36, and 7.37. It consists of all necessary parts of the mechanical refinement of biomass (crushing, sorting, drying, humidifying, and cooling) as well as the compaction technology (pelleting and briquetting) to be used. This experimental line is currently used to investigate pellet pressing using a flat disk die with vertical axis of rotation. The line is also available with hydraulic and mechanical (screw press) briquette pressing as a consequence of the scientific research performed at our institute. Machines present in this line can be used for the experimental verification of theoretical results.

Fig. 7.35 Part of the laboratory production line

Fig. 7.36 Refinement stage

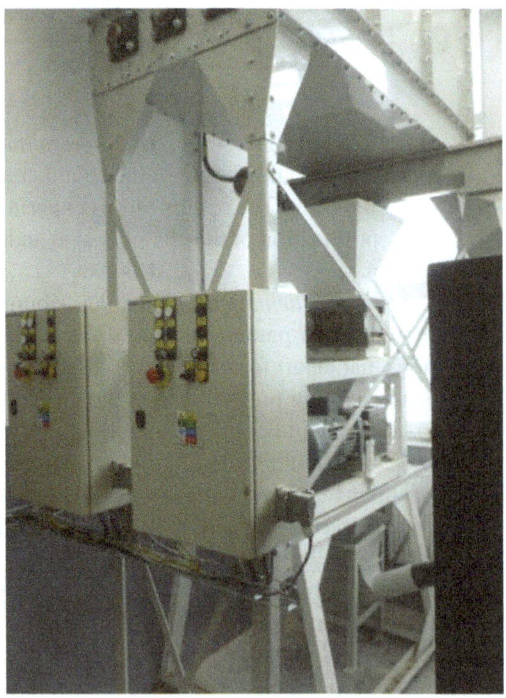

Fig. 7.37 KAHL pellet mill

References

1. Križan, P., Šooš, Ľ., Matúš, M.: Optimalisation of briquetting machine pressing chamber geometry. In: Machine Design, pp. 19–24 (2010). ISSN 1821-1259
2. Matúš, M., Križan, P.: Influence of structural parameters in compacting process on quality of biomass pressing. Aplimat J. Appl. Math. **3**(3), 87–96 (2010). ISSN 1337-6365
3. Križan, P., Vukelič, D.: Shape of pressing chamber for wood biomas compacting. In: Quality Festival 2008: 2nd International Quality Conference. Kragujevac, May 13–15, 2008. University in Kragujevac, Kragujevac (2008). ISBN 978-86-86663-25-2
4. Križan, P.: The Process of Pressing Wood Waste, 1. vyd, 197 p. Publishing House STU, Bratislava (2014). ISBN 978-80-227-4251-1 (in Slovak)
5. Križan, P., Matúš, M.: Expert analysis of Compaction of Sunflower Husks, 1st ed., 9 p. Bratislava STU, Faculty of Mechanical Engineering (2016) (in Slovak)
6. Križan, P., Matúš, M.: Expert Analysis of Cocoa Shell Compaction, 1st ed., 14 p. Bratislava STU, Faculty of Mechanical Engineering (2016) (in Slovak)
7. Bábics, J., Ondruška, J., Šooš, Ľ., Križan, P., Matúš, M.: Development of a Dry Sludge Outlet Pressure Seal Against Steam Leakage, 1st ed., 24 p. Bratislava STU, Faculty of Mechanical Engineering (2017) (in Slovak)
8. Šooš, Ľ., Pokusová, M., Križan, P., Beniak, J., Matúš, M.: Expert Analysis of Sludge from Waste Water Cleaners Recovery, 11 p. STU in Bratislava FME, Bratislava (2013) (in Slovak)
9. Šooš, Ľ., Kolláth, Ľ., Krížik, P.: Treatment of sewage sludge for the purpose of their energy recovery. In: TOP 2005. Proceedings of International Conference, pp. 139–148. STU in Bratislava, Bratislava (2005). ISBN 80-227-2249-9 (in Slovak)
10. Šooš, Ľ., Križan, P., Matúš, M., Ondruška, J., Bábics, J.: Research of dry sludge compaction system, 25 p. Bratislava STU Faculty of Mechanical Engineering (2018) (in Slovak)

11. Kers, J., Križan, P. Letko, M., Šooš, Ľ., Kask, Ü. Gregor, A.: Mechanical recycling of compounded polymeric waste and evaluation of briquetting parameters. In: Proceedings of the 7th International Conference of DAAAM Baltic Industrial Engineering: Tallinn, Estonia 22–24 April 2010, pp. 468–473. Tallinn University of Technology, Tallinn (2010). ISBN 978-9985-59-982-2
12. Križan, P., Matúš, M., Šooš, Ľ., Kers, J., Peetsalu, P., Kask, Ü., Menind, A.: Briquetting of municipal solid waste by different technologies in order to evaluate its quality and properties. Agron. Res. **9**(Spec.iss. 1), 115–123 (2011). ISSN 1406-894X
13. Šooš, Ľ., Kolláth, Ľ., Onderová, I., Križan, P., Ondruška, J. Čačko, V. Biath, P.: The Concept of a Separation Line for the Separation of Municipal Waste, 103 p. Bratislava STU Faculty of Mechanical Engineering (2014) (in Slovak)
14. Matúš, M., Križan, P., Šooš, Ľ.: Design of Technologies for Transport of Fuels from Non-wood Biomass, 1st. ed., 29 p. Bratislava STU, Faculty of Mechanical Engineering (2017) (in Slovak)
15. Matúš, M., Šooš, Ľ., Ondruška, J., Biath, P., Čačko, V.: Expert Analysis of Digestate Pelleting from Biogas Plants, 1st. ed., 10 p. Bratislava Faculty of Mechanical Engineering STU (2015) (in Slovak)

Index

P. Križan, *Biomass Compaction*, https://doi.org/10.1007/978-3-030-89956-1

The manufacturer's authorised representative in the EU is Springer Nature Customer Service Centre GmbH, Europaplatz 3, 69115 Heidelberg, Germany. If you have any concerns regarding our products, please contact ProductSafety@springernature.com

Printed and bound by CPI Group (UK) Ltd, Croydon, CR0 4YY
28/09/2026
02224089-0019